Disrupting Tradition: Research and Practice Pathways in Mathematics Education

EDITORS

William F. Tate
Washington University in St. Louis, Missouri

Karen D. King
New York University, New York, New York

Celia Rousseau Anderson
University of Memphis, Tennessee

NATIONAL COUNCIL OF
TEACHERS OF MATHEMA....

D0813508

Copyright © 2011 by
THE NATIONAL COUNCIL OF TEACHERS OF MATHEMATICS, INC.
1906 Association Drive, Reston, VA 20191-1502
(703) 620-9840; (800) 235-7566; www.nctm.org
All rights reserved

Library of Congress Cataloging-in-Publication Data
Disrupting tradition : pathways for research and practice in mathematics
education / William F. Tate, Karen D. King, Celia Rousseau Anderson,
editors.
    p. cm.
  Includes bibliographical references.
  ISBN 978-0-87353-636-3
  1. Mathematics--Study and teaching--United States. 2. Curriculum
planning--United States. I. Tate, William F. II. King, Karen D. III.
Anderson, Celia Rousseau.
  QA13.P38 2010
  510.71--dc22
                          2010051546

The National Council of Teachers of Mathematics is a public voice of mathematics education, supporting teachers to ensure equitable mathematics learning of the highest quality for all students through vision, leadership, professional development, and research.

Printed in the United States of America

Table of Contents

Introduction

Linking Research and Practice in Mathematics Education

Over the past two decades, numerous initiatives have encouraged partnerships among researchers and mathematics education professionals (e.g., teachers, specialists, math coordinators, and so on). The National Council of Teachers of Mathematics (NCTM) has recently implemented a strategically focused initiative on linking research and practice (Heid et al. 2006). Many of these types of initiatives and related position papers aim to develop coherent, interdependent, research-and-practice strategies. What forms have these research–and-practice activities taken? What lessons have been learned from these collaborations? This volume examines both these questions.

Examining these questions first requires recounting some history. The annals of research have traditionally linked with efforts to understand better the quality factors associated with curriculum, instruction, and learning conditions. The story of the research enterprise in the field of mathematics education, however—particularly concerning building bridges to the school leadership and educational policy communities—can be described as *evolving*. In fact, mathematics education's focuses as a field were, for many years, very specified and largely decoupled: some scholars focused on learning, whereas others examined teaching or curriculum. However, practitioners have to negotiate multiple influences on students' mathematical proficiencies. In contrast, Romberg and Carpenter (1986) argued the need in mathematics education for a combined, synthesized concept of two distinct areas of research—research on teaching and research on learning. Their review implied that separating these two areas works against establishing a well-informed understanding of the developmental processes associated with mathematical competency.

Research on learning suggests that the cognitive demand of mathematical tasks links to students' understandings and skills. It follows that the aim to support learning associates with curriculum choices and teachers' action. Mathematical tasks are embedded in instructional materials that link to curriculum management tools. These tools are interpreted and implemented to varying degrees by teachers who have a range of talents and understandings. Research on teaching, however, indicates that teachers with high expectations, who provide help and support, positively influence students' outcomes. High expectations include furnishing developmentally

William F. Tate's research and development has been funded by the National Science Foundation under Award No. ESI0227619. Any opinions, findings, and conclusions or recommendations expressed here are those of the author and do not necessarily reflect the views of the National Science Foundation.

appropriate and cognitively demanding mathematical tasks. The interdependence among teaching, learning, and curriculum as part of the design of an effective instructional regime appears greatly significant to aims to link research and practice. Yet, recognizing that significance is only one step in the strategy to create the link.

A basic understanding of the school environment suggests that school leaders are central to the design, maintenance, and improvement of instructional practice. Rowan (1995) called for a research agenda that linked learning, teaching, and educational administration. In his opinion, this bridge would address a criticism of research on educational administration—more specifically, that scholarship in the field of educational administration has failed to address issues of teaching and learning in schools. Although he directed his message largely to colleagues in educational administration, he could have as easily aimed it at the mathematics education community. Mathematics education's idea of a robust research-and-development effort that examines important links among teaching, learning, and leadership was underdeveloped. Fortunately, this void in the literature has not been ignored. Stein and Nelson (2003) argued that leadership content knowledge is a missing paradigm in education research. They described leadership content knowledge as a knowledge of school subjects (e.g., mathematics) and of how students learn the content. Stein and Nelson's analysis implicitly stated that leaders require a sound understanding of teaching, learning, curriculum, and other factors influencing students' skill development and understanding. A National Research Council report (Elmore and Rothman 1999, p. 3) made this argument explicitly:

> The theory of action behind an education improvement system relies on information and responsibility. Everyone in the system—students, parents, teachers, administrators, and policy makers at every level—needs high-quality information about the quality of instruction and student performance. At the same time, everyone needs to be responsible for fulfilling his or her role improving results. The key is transparency: everyone should know what is expected, what they will be measured on, and what results imply for what they should do next.

This report focused largely on building accountability and assessment systems to support standards-based reform. Henig and colleagues (1999) warned that many accountability models almost exclusively rely on test scores, mechanically, to identify successful and unsuccessful schools. They noted that this approach encourages centralized authorities to design and implement reforms but does little to make those reforms work. We submit that important parts of a system's capacity are (1) developing sustained engagement among school professionals, researchers, and policymakers and (2) making crucial factors influencing students' outcomes in mathematics central to the collaborative process. Focusing strictly on students' outcomes is not enough. Understanding, and intervening on, factors that influence students' understanding and skills in mathematics are vitally important, too. One strategy for intervening is to bridge the research-and-practice divide by forming alliances that include both scholars and practitioners. This kind of alliance has great potential, but it also has a problem; no grand theories or how-to books focus on building sustained, research-based practice in mathematics on a scale that would make a difference. Enter the colleagues who were invited to participate in this book.

The chapters in this book offer (1) insights into the role research can play in supporting and improving school mathematics and (2) specific examples of that role, while raising questions that are valuable for the mathematics education community. NCTM has sponsored the development of two handbooks aimed to serve the research community. We do not intend to replicate these two documents. Instead, our goals are (1) to illustrate where strategic partnerships have linked research findings to the design of practice and programmatic endeavors and (2), in many instances, to illustrate generating evidence to guide both educational decision making and routines modification related to school mathematics.

Beginning with James P. Spillane, several authors in this volume examine how research can influence practice, policy, and leadership in mathematics teaching and learning. Like the other authors of the first five chapters, Spillane demonstrates how tool development is an important aspect of the research-and-practice integration process. Drawing on theoretical and empirical research in distributed cognition and sociocultural activity theory, he describes a hypotheses-generating research-and-development (R & D) effort, That effort is part of the Distributed Leadership Studies (DLS), where knowledge of the *how* of leading and managing is driving the R & D program. Spillane states the four major components of DLS: (1) designing and validating research or diagnostic instruments for gathering policy relevant data, (2) analyzing leadership and management arrangements in schools, (3) engaging district policymakers and school practitioners with research findings related to their own schools, and (4) designing curriculum modules that engage school personnel distributively in diagnosis and design. Three examples of DLS collaborations illustrate the resulting research-to-practice interactions.

In chapter 2, "Surveys of Enacted Curriculum and the Council of Chief State School Officers Collaborative," Andrew Porter and Jennifer McMaken explain the value of the Surveys of Enacted Curriculum as a part of a practical, reliable reporting tool focused on instructional practice and content being taught in classrooms. The chapter describes a collaboration among the Wisconsin Center for Educational Research, the Council of Chief State School Officers, state education agencies, and service provider organizations. This collaboration helps state agencies and local school districts implement data-collection procedures and reporting tools that generate evidence on how instructional practice and content align with required mathematics standards and assessments of students' learning. The collaboration also provides professional development to state and local educators on uses of enacted curriculum data as part of decision making in instructional improvement.

In chapter 3, "The Role of Tools in Bridging Research and Practice in an Instructional Improvement Effort," Mary Kay Stein, Jennifer Russell, and Margaret Schwann Smith depict a collaborative project involving university-based researchers and a grades 6–10 public school. The project examined the role of technology-enhanced lesson planning in guiding schoolwide improvement. They argue that school districts will not realize the promise of tools as supports for large-scale, research-based improvement without proper attention to teachers' learning and organizational reinforcement. They define tools as artifacts (e.g., curriculum materials, computer-based programs, observation protocols, and rubrics) that apply insights from scholarship directly and readily to practice. The authors describe how tools can be embedded in a project-based research effort, where insights generated from the collaboration both influence the research literature and link to teachers' planning, enactment, and reflection. Also, they demonstrate how to align lesson-planning routines with the work of school leadership. The preliminary findings from this project are promising.

In chapter 4, "Building Bridges between Research and the Worlds of Policy and Practice:

Lessons Learned from PROM/SE," William H. Schmidt paints a picture of how reports consisting of modified instrumentation from the Third International Mathematics and Science Study supported the development of capacity-building activities and intervention design. The chapter looks into how research guided policy decisions at various levels of the educational system—school, district, and state. Schmidt voices a warning common throughout the volume: he argues that the collaboration level required to generate meaningful research-and-practice exchanges is difficult when resistance threatens the process. He offers several important lessons learned from collaborations with scores of school districts.

In chapter 5, "Teachers' Use of Standards-Based Instructional Materials: Partnering to Research Urban Mathematics Education Reform," Karen D. King passes on lessons that emerged from a project designed to study how middle grades teachers use standards-based mathematics instructional materials in a reform effort in an urban school district. Like the authors of the first four chapters, King is intent on discovering the nature and quality of the instructional realities that teachers encounter in implementing the materials. In studying survey results, including information gleaned from the Surveys of Enacted Curriculum (see Porter and McMaken 2010); she found that teachers' actions diverged significantly from those the district envisioned and articulated in the curriculum pacing guide. The chapter concludes with observations and cautions for those interested in research, policy, and practice concerning collaborations.

In chapter 6, "Examining What We Know for Sure: Tracking in Middle Grades Mathematics," Lee V. Stiff, Janet L. Johnson, and Patrick Akos recount research findings from an effort to document the influence of curriculum differentiation practices and policies on opportunity to learn and students' related outcomes. The authors used a variety of methodological approaches and data sources to guide a partnership of counselors, mathematics teachers, and administrators. The findings reflect unintended consequences of existing placement policy and educators' decision making. The implications for traditionally underserved students are significant.

In chapter 7, "Mathematics Education, Language Policy, and English Language Learners," Marta Civil looks closely into the complex world of language ideology and the opportunity to learn mathematics. She uses evidence gathered by the Center for the Mathematics Education of Latinos/as (CEMELA). Relying on sociocultural approaches and qualitative data, Civil explores mathematics placement policy and parents' perceptions of their children's mathematics education. Civil also describes classroom conversations between teachers and students, which offer a glimpse into the intricate relationship between language and mathematics learning. Guided by her research, Civil posits that Arizona's language policy has in some instances negatively influenced mathematics discourse and learning. She recommends a better integration of research, policy, and practice to deal with that negative influence.

In chapter 8, "Elementary Mathematics Specialists: A Merger of Policy, Practice, and Research," Patricia F. Campbell communicates details about a Virginia collaborative project involving four universities and five school districts organized to develop coursework with two aims. The first aim was to produce an integrated academic experience where mathematics content and pedagogical content knowledge were central to the coursework pathway. The second aim was to develop a coherent leadership-coaching experience where the courses shaped the participants' (e.g., in-service teachers) understanding of current research on mathematics teaching and learning and supported their growth as change agents or coaches. In tandem, the project's aims were part of a concerted human capital development strategy focused on the production of elementary mathematics specialists charged to support the improvement of mathematics teaching and learning in school settings. Research was a part of the collaborative's core activities. For example, the coursework development process relied on fundamental understandings gleaned from scholarship in education and related literatures. Also, the collaborative's efforts included a randomized control trial designed to evaluate the effect of mathematics specialists on students' achievement and teachers' beliefs. The research findings influenced legislative and regulatory discussions as well as local school policy developments. (For further reading on the development of mathematics instructional specialists, see Martin et al. 2010.)

In chapter 9, "Transforming East Alabama Mathematics (TEAM-Math): Promoting Systemic Change in Schools and Universities," W. Gary Martin, Marilyn E. Strutchens, Stephan Stuckwisch, and Mohammed Qazi convey essential elements of their National Science Foundation–funded math and science partnership (MSP) located in east Alabama. The MSP, referred to as TEAM-Math, included two universities and fourteen school districts. Research guided the design of the partnership's leadership model, professional development activities, and community and parental engagement strategies. Research by two authors of other chapters in this volume (Campbell 2010; Spillane 2010) shaped the MSP's conceptual underpinnings. TEAM-Math borrowed from the research study of Campbell and colleagues (2003) regarding the effectiveness of mathematics specialists as school-based supports. In addition, Spillane's (2000) distributed leadership model guided the thinking associated with TEAM-Math's organizational structure. Preliminary TEAM-Math findings suggest the partnership is moving toward the development of important lessons related to teachers' quality and students' learning.

In chapter 10, "The SCALE Project: Field Notes on a Mathematics Reform Effort," Terrence Millar and Mathew D. Felton chronicle more than decade of collaborative work organized to improve grades K–16 mathematics education. One major focus of their remarks is the cultural and institutional challenges associated with multiorganizational partnerships as well as internal institutional dilemmas. Their chapter is a reminder that collaboration as a process is worthy of study, especially if linking research and practice in mathematics education is an important objective. Collaborations are not natural acts; instead, the process includes difficult decisions, comprise, and communication strategies. Using an insider lens, the authors share lessons learned from years of partnership work. The lessons discussed in this chapter are relevant, because all this book's chapters involve some form of collaboration or partnership.

Finally, "Reflection" by William F. Tate offers a few concluding remarks about the future of research-and-practice collaborations in mathematics education. He argues that research-and-practice collaborations should be a standard regime in movements to improve mathematics teaching and learning.

A Cypriot proverb states, "We must convince by reason, not prescribe by tradition." The chapters in this book describe efforts to bring reason to the mathematics education field by way of purposeful links between disciplined inquiry and practice. As a collective, the authors raise questions about why so many colleagues are comfortable with traditional practice in mathematics education. Moreover, the chapters offer various research and practice pathways to disrupt tradition in mathematics education. Why disrupt tradition? Although tradition brings a level of security, it also can foster stagnation and decline. This book aims, where appropriate, to disrupt traditional customs, folkways, and thinking by providing instances of pathways that make serious attempts to promote mutually informing, disciplined inquiry and practice in mathematics education.

REFERENCES

Campbell, Patricia. "Elementary Mathematics Specialists: A Merger of Policy, Practice, and Research. " In *Disrupting Tradition: Research and Practice Pathways in Mathematics Education,* edited by William F. Tate, Karen D. King, and Celia Rousseau Anderson, pp. 93–103. Reston, Va.: National Council of Teachers of Mathematics, 2010.

Campbell, Patricia, Andrea Bowden, Steve Kramer, and Mary Yakimowski. "Mathematics and Reasoning Skills (No. ESI 9554186)." College Park, Md.: University of Maryland, MARS Project, 2003.

Elmore, Richard F., and Robert Rothman, eds. *Testing, Teaching, and Learning: A Guide for States and School Districts.* Washington, D.C.: National Academies Press, 1999.

Heid, M. Kathleen, James A. Middleton, Matthew Larson, Eric Gutstein, James T. Fey, Karen King, Marilyn E. Strutchens, and Harry Tunis. "The Challenge of Linking Research and Practice." *Journal for Research in Mathematics Education* 37 (January 2006): 11.

Henig, Jeffery R., Richard C. Hula, Marion Orr, and Desiree Pedescleaux. *The Color of School Reform: Race, Politics, and the Challenge of Urban Education.* Princeton: Princeton University Press, 1999.

Martin, W. Gary, Marilyn E. Strutchens, Stephen Stuckwisch, and Mohammed Qazi. "Transforming East Alabama Mathematics (TEAM-Math): Promoting Systemic Change in Schools and Universities." In *Disrupting Tradition: Research and Practice Pathways in Mathematics Education,* edited by William F. Tate, Karen D. King, and Celia Rousseau Anderson, pp. 105–18. Reston, Va.: National Council of Teachers of Mathematics, 2010.

Porter, Andrew C., and Jennifer McMaken. "Surveys of Enacted Curriculum and the Council of Chief State School Officers Collaboration." In *Disrupting Tradition: Research and Practice Pathways in Mathematics Education,* edited by William F. Tate, Karen D. King, and Celia Rousseau Anderson, pp. 21–31. Reston, Va.: National Council of Teachers of Mathematics, 2010.

Romberg, Thomas A., and Thomas P. Carpenter. "Research on Teaching and Learning Mathematics: Two Disciplines of Scientific Inquiry," In *Handbook of Research on Teaching,* 3rd ed., edited by Merlin C. Wittrock, pp. 850–73. New York: Macmillan, 1986.

Rowan, Brian. "Learning, Teaching, and Educational Administration: Toward a Research Agenda." *Educational Administration Quarterly* 31 (August 1995): 344–54.

Spillane, James P. "Cognition and Policy Implementation: District Policymakers and the Reform of Mathematics Education." *Cognition and Instruction* 18 (June 2000): 141–79.

———. "The Distributed Leadership Studies: A Case Study of Research *in* and *for* School Practice." In *Disrupting Tradition: Research and Practice Pathways in Mathematics Education,* edited by William F. Tate, Karen D. King, and Celia Rousseau Anderson, pp. 7–19. Reston, Va.: National Council of Teachers of Mathematics, 2010.

Stein, Mary K., and Barbara S. Nelson. "Leadership Content Knowledge." *Educational Evaluation and Policy Analysis* 25 (Winter 2005): 423–48.

The Distributed Leadership Studies: A Case Study of Research *in* and *for* School Practice

James P. Spillane

Over the last decade, researchers in the Distributed Leadership Studies (DLS) at Northwestern University have developed a framework for examining school leadership and management with an emphasis on their relations to classroom instruction. (For more information on DLS, see www.distributedleadership.org.) Drawing on theoretical and empirical work in distributed cognition and sociocultural activity theory, we have built our distributed perspective around two aspects—principal plus and practice (Spillane 2006; Spillane, Halverson, and Diamond 2001, 2004). The principal-plus aspect acknowledges that many people contribute to the work of leading and managing schools. The practice aspect describes the practice of leading and managing, framing this practice as emerging from the interactions among school leaders and followers, mediated by the situation in which the work occurs. Practice is more about interaction than action. The school subject matter—mathematics, science, and language arts—has figured prominently in our efforts to build knowledge about and for the practice of leading and managing.

With respect to mathematics, attention to the school subject is pertinent for at least two reasons. First, high school teachers' views of their subjects differ on dimensions such as the scope of the subject, the degree to which the material is sequenced, whether the subject is static or dynamic, and the degree to which the subject is core or peripheral (Grossman and Stodolsky 1995). For example, mathematics teachers are more likely than teachers of other subjects to see their work as routine (Rowan, Raudenbush, and Cheong 1993), highly sequenced, and static (Grossman and Stodolsky 1995). At the elementary school level the subject also mattered. The topics, sequence of instruction, and intellectual goals of mathematics, compared with those of social studies, were more uniform across different fifth-grade teachers' classrooms (Stodolsky 1988). Also, evidence suggests that elementary school language arts teachers' conceptions of themselves as teachers and as learners about teaching differ from those of elementary school mathematics teachers. This difference influences how the teachers construct and respond to reform efforts as well as what learning opportunities they seek (Drake, Spillane, and Hufferd-Ackles 2001; Spillane 2005). Hence, we would expect that efforts to lead and manage mathematics instruction might differ from efforts to lead and manage instruction in other school subjects at both the elementary and secondary school levels. Second, some evidence suggests that elementary schools marshal fewer resources to support reform in mathematics than they do for language arts: most schools have a full- or part-time reading coordinator, often supported with Title I monies (Price and Ball 1997). As a result, the resources available for leading and managing instruction may differ depending on the school subject.

This chapter uses our hypotheses-generating research-and-development work, part of the DLS, as an example of connecting research with practice and policy about mathematics education. It begins by describing our research-and-development work on school practice and our various goals. This chapter then describes some ways in which the DLS have forged connections with policymakers and practitioners through three different partnering experiences. We next consider, in more detail, a facet of our work that uses our research findings to engage policymakers and practitioners in diagnosing and design work to develop practical knowledge—*how* knowledge, as is distinct from *what* knowledge. Finally, this chapter concludes by reflecting on some of the challenges the DLS have encountered.

The DLS Research and Development Program

In our use, a *distributed perspective* is not a blueprint for leading and managing, but rather, a framework for researchers and practitioners to use in (1) diagnosing the practice of leading and managing and (2) designing for its improvement (Spillane 2006; Spillane and Diamond 2007). Keeping with this definition of the term, the DLS are committed to developing knowledge about leading and managing, especially knowledge for practice—knowledge of the *how* of leading and managing. Although a sizable knowledge base exists about the *what* of leading and managing, we know less about the *how*—the practical knowledge that school leaders use in their day-to-day practice. For example, research informs us that monitoring instruction is important for instructional innovation and improvement (Firestone 1989). Still, the available knowledge base has little to say about the *how* of monitoring instruction in general and in mathematics instruction in particular. Without a rich understanding of the *how*, policymakers and researchers have difficulty contributing to improving school leadership and management. More important, some evidence suggests that these challenges are subject-matter-specific. For example, some recent work suggests that, concerning school leaders' thinking about their work, the school subject does, indeed, matter (Burch and Spillane 2003). School leaders' cognitive scripts for the work of instructional improvement differed depending on the subject matter. Whereas most school leaders viewed the expertise for leading change in language arts instruction as homegrown, inside the schoolhouse, most saw the expertise for leading change in mathematics coming from outside the school.

One component of DLS's work involves designing and validating research or diagnostic instruments, such as logs of practice and social network tools (Camburn, Spillane, and Sebastian, under review; Spillane and Zuberi 2009; Pitts and Spillane 2009; Pustejovsky et al. 2009; Pustejovsky and Spillane 2009). Although these instruments are designed for gathering data, policymakers and practitioners can also use them to generate data that support reflection in and on the practice of leading and managing. We also work directly with schools and districts in our studies to share research findings so that they can give their policymakers and practitioners the opportunity to reflect on practice using data from their own schools.

A second component of our work described and analyzed leadership and management arrangements in schools (Spillane and Diamond 2007; Spillane, Camburn, and Pareja 2007; Spillane, Hallett, and Diamond 2003). For example, our research showed that leading and managing instruction involved not only the principal, but also a host of others—assistant principals, curriculum specialists, mentor teachers, and department chairs (Spillane 2006; Spillane and Diamond 2007; Spillane, Hunt, and Healey 2009). Moreover, leadership and management arrangements in elementary schools differed depending on the school subject (Spillane 2005, 2006). This component of the work explored relations among those arrangements, organizational conditions, and instructional innovation. By using mixed research methods, the work focused on relations among school leadership, school management, and instructional improvement. Although other researchers were the primary audience, we also worked to engage policymakers and practitioners indirectly and directly with our findings.

A third component of the DLS's work, especially crucial to developing knowledge for practice, engaged district policymakers and school practitioners who participated in our studies with research findings for their own schools. Specifically, we compiled reports for individual schools and then conducted workshops that focused on the findings in the individualized reports.

A fourth component designed curriculum modules that engaged school staff in diagnostic and design work using the distributed perspective. These modules featured our research on school leadership and management extensively. First, we used cases from our work to engage participants in understanding what taking a distributed perspective to school leadership and management entails (Spillane 2006; Spillane and Diamond 2007; Spillane and Coldren, in progress). Second, we used our research findings to engage participants in diagnostic work from a distributed perspective. Third, we used our research and diagnostic instruments to help participants transfer the findings from our empirical case studies to their own schools.

Our work in the DLS has been made possible by establishing connections with district policymakers, school practitioners, and other colleagues engaged in research and development. The next section describes these connections by exploring how they have facilitated our efforts and by identifying similarities and differences in our partnerships.

Connecting Researchers, Policymakers, and Practitioners

Over the past decade the DLS have worked with several partners to implement research and development. Some of these partnerships have connected directly to policymakers and practitioners, whereas others have connected indirectly, mediated by other research-and-development projects. The three examples below—Chicago Public Schools (CPS), Math in the Middle (M^2), and the Penn Center for Educational Leadership—illuminate the different ways in which the DLS have forged connections to policy and practice.

Chicago Public Schools (CPS)

Our work with CPS's Office of Mathematics and Science has taken various forms, all directly connecting to practitioners and district policymakers. Some CPS elementary schools, grades K–8, were the original study sites for DLS data collection, starting in 1999. (This work's funding came from the Spencer Foundation grant #200000039 and the National Science Foundation's [NSF] grant #REC-9873583. James Spillane was principal investigator.) We negotiated these partnerships, which consisted of conventional researcher-practitioner relations for data collection, with individual schools. Further, with support from the Carnegie Corporation of New York, we continued our work at other CPS schools, developing teaching modules that engaged their school leaders in the same diagnosis and design of leading and managing. (Penelope Peterson and James Spillane served as coprincipal investigators.) These modules used findings from our earlier research as well as case studies we developed from data analysis for teaching purposes. We used the principles of design research, engaging in a process of progressive refinement, in which we tested and refined our modules on the basis of results of prior pilot studies (Collins, Joseph, and Bielaczyc 2004). After we piloted each module with teams of school leaders from two CPS schools, the leaders participated in focus groups, conducted by an independent researcher, that reflected on particular units. From feedback from the focus groups, the instructor's notes on the session, and two project researchers who observed each session, we revised the modules. Using the modules with other schools in other districts, we continued to redesign on the basis of feedback from participants while keeping with our distributed perspective on leading and managing.

Also, with funding from the NSF's RETA program (RETA EHR-0412510), we piloted multiple prototypes of the School Staff Social Network Questionnaire (SSSNQ) and the Leadership Daily Practice log in a purposeful sample of CPS elementary and middle schools (Spillane and Zuberi 2009; Pitts and Spillane 2009; Pustejovsky et al. .in press; Pustejovsky and Spillane 2009). The SSSNQ, a Web-based survey instrument, collected data about leadership and management arrangements for mathematics and other subjects in elementary and middle schools (see figs. 1.1, 1.2, and 1.3). For example, the SSSNQ used social network items to gather data about the advice and information networks of school staff concerning core school subjects. By conceiving of leadership as social influence relations about instruction, the SSSNQ used a social network approach to measure leadership interactions.

Next, we used the SSSNQ with twenty-three Chicago public schools—both grades K–8 and middle schools—that were participating in the Cluster 4 Middle Grades Program (C4MGP), a leadership and school-restructuring initiative. Here our partnering arrangements shifted from working with individual schools directly to working with schools through a district-sponsored initiative. Our work had two distinct components. First, funded by the Searle Funds at the Chicago Community Trust (grant #C2006-01385), the CPS Office of Mathematics and Science used our modules in a nine-month training program for 23 school principals and their area instructional officers. One of the modules' three components addressed the leadership part of C4MGP professional development; the other two focused on mathematics and language arts, respectively. Our leadership modules pressed participants to pay careful attention to the unique challenges of leading and managing improvement in mathematics education by taking the school subject seriously. Second, we invited all 23 schools to have their staff

Fig. 1.1. Screen shot from SSSNQ, version 2—math advice questions, page 1

Fig. 1.2. Screen shot from SSSNQ, version 2—math advice questions, page 2

Fig. 1.3. Screen shot from SSSNQ, version 2—math advice questions, page 3

take the SSSNQ instrument as part of C4MGP. In spring 2007, we administered the survey to the nineteen schools that accepted our invitation. In June 2008, we administered the SSSNQ again, though only to those twelve schools whose response rate in 2007 was more than 70 percent. We repeated the data collection with a still smaller sample in spring 2009, with schools with high response rates from 2008.

After each data-collection period, the schools with response rates of more than 80 percent received (1) individualized reports based on our analysis of the data and (2) a workshop designed to engage study participants in interpreting the data for their school.

For example, we conducted, on the basis of results from DLS analysis of the 2007 and 2008 SSSNQ data, a three-hour workshop in December 2008 for school principals and teacher leaders from participating schools. Besides engaging participants with data from their own schools, the workshop had them use data from one particular school to identify patterns of change from one school year to the next. At the end of the workshop, participants received a homework assignment to use the advice-and-information network about mathematics teaching from their school to (1) identify shifts in interaction patterns over time and (2) consider the implications of these changes for their efforts to improve mathematics education. Specifically, each leadership team was told to analyze carefully the mathematics advice network sociograms for both school years at their school. The team would then identify pivotal changes in the mathematics advice network from one year to the next, clearly stating why these changes were important to their school's improvement of mathematics instruction. Next, a team would identify the change in the network structure that they considered most negative toward improving mathematics instruction. Each team then developed a plan that clearly specified a theory of action to remedy the problem. They also entertained reasons why their proposed solution might not work. Teams worked with their school network data and other data sources (e.g., staff mobility, changes in formally designated leadership positions, changes in grade-level assignments) to complete this assignment. In February 2009, each team presented their school with (1) a diagnosis of the situation based on their interpretation of the data and (2) their prognosis for improvement. Thus, working with actual data from the teams' own schools, this assignment engaged the teams in what the DLS demonstrated: diagnosing a problem at their schools related to mathematics education, and designing a remedy for that problem.

Math in the Middle (M²)

The DLS collaborated with professors James Lewis and Ruth Heaton, codirectors of the M² Institute Partnership at the University of Nebraska at Lincoln (UNL), in 2006 to collect data about leading mathematics instruction in middle schools. Funded by a math-science partnership grant from the NSF's RETA program (grant EHR-0142502), and with additional support from UNL's Math and Science Teachers for the 21st Century Program of Excellence, M² offered a twenty-five-month master's degree program for outstanding middle school math teachers who became M² associates. The program focused on developing strong intellectual leaders for middle school mathematics by providing participants opportunities to develop their knowledge and skills in mathematics, pedagogy, and leadership. These M² associates would ideally serve as leaders for mathematics education in their schools, school districts, and educational service units.

Such leadership work might include providing colleagues professional development and offering them guidance about mathematics content or pedagogy.

Working with M² researchers, we tailored our SSSNQ so that we could gather data relevant to their efforts to improve leadership for middle school mathematics. Our work here centered on understanding *whether*, rather than *how*, M² associates were primary advice givers concerning mathematics education in their schools. In other words, did M² associates assume the leadership roles for mathematics that the M² Institute Partnership prepared them for?

Starting in spring 2006, M² administered the SSSNQ annually for three consecutive years to 96 teacher associates across three cohorts that had participated in the M² program since its inception. They also administered the SSSNQ to teachers in all ten middle schools in Lincoln, Nebraska, in 2007, and again in 2008, in order to investigate the roles of the M² teacher associates in their school setting. The longitudinal data enabled us to analyze changes in formal and informal leadership for mathematics education in these schools over time, particularly the roles of the M² associates.

Comparing the subset of M² associates—between one and four in each middle school—to other teachers with similar leadership roles gave us insight into how M² associates act as teacher leaders. For example, M² associates in the ten Lincoln middle schools reported more sources of advice from outside their school buildings than other teachers in similar roles. M² associates who were math teachers listed an average of 2.1 external advisors in the 2007 survey, whereas other math teachers listed an average of 0.7 (Pustejovsky et al. 2009). Research has demonstrated that access to information from outside an organization's boundary benefits innovation and productivity (Burt 2000; Coburn and Russell 2008; Reagans and McEvily 2003). Even more interesting, people tended to name M² associates as advisors more often than they did other colleagues. The 2007 survey found an average of 8.8 colleagues named M² associate math teachers as advisors, whereas an average of 7.0 colleagues named other math teachers thus. Because of the absence of baseline data, we could not draw causal claims about relations between the M² program participation and leadership arrangements for mathematics education. Our analysis did suggest, however, that M² associates were behaving as the program's designers intended, as primary advice givers about mathematics education in their schools. Our analysis further suggested that M² associates both drew on and contributed to an advice network, the boundary of which participation in the M² program defined. This characteristic was important because work on organizations in general, and on schools in particular, suggested that ties extending beyond organizational boundaries are crucial to innovation. They ensure the flow of new knowledge into an organization and help guard against "group think," (i.e., abandoning individual thought in favor of blindly following group consensus).

Moving beyond *what* knowledge to engage study participants in developing and articulating *how* knowledge, we generated individualized reports for each Lincoln middle school based on our analysis of the data compiled from the SSSNQ. We also conducted presentations to engage the middle school principals with the data from their school. The results of our work here resembled the individualized reports written for CPS schools. Our primary purpose was to engage practitioners in diagnostic work from a distributed perspective using data from their own schools.

Penn Center for Educational Leadership

Funded by the Annenberg Foundation, the DLS have worked with John DeFlaminis and Jim O'Toole of the Penn Center for Educational Leadership to design a distributed-perspective leadership development program for teams of leaders from schools in the School District of Philadelphia (SDP). An essential component of this work has used our DLS leadership development modules with multiple cohorts of SDP elementary, middle, and high school leaders. Although this effort did not focus on mathematics education, mathematics education was a primary component. Throughout the modules, we pressed school leaders to think about how the school subject affected leading and managing school instruction. For example, one core activity in our modules required school teams to analyze social network data for advice in both mathematics and language arts from the same school, in order to identify similarities and differences between the two networks. Participants then developed working hypotheses about the differences they identified.

The Penn Center for Educational Leadership has mediated all our work with SDP; DLS modules were only two of several that their program used. SDP, however, incorporated the distributed perspective on leading and managing throughout its curriculum.

Research for Practice: Priming the Development of *How* Knowledge

A vitally important feature of the DLS is our effort to connect with policy and practice not only through the conventional means of generating research findings—*what* knowledge—but also by creating tools that enable practitioners to reflect on their practice and to develop, and make explicit, practical knowledge—*how* knowledge—about leading and managing. The DLS modules are one obvious means to that end. The modules use data from our work to engage practitioners in diagnostic and design work using a distributed perspective. We employ the "case study" approach frequently in the modules and present the empirically grounded cases in various forms. One prominent form is written narratives generated from data from a particular school and focused on a particular issue (e.g., diagnosing problems in practice, using students' achievement data in decision making).

Social network diagrams or maps based on data from real schools are another form of case study in the modules. This form is especially useful in engendering dialogue among participants about practice (see fig. 1.4). We have school principals or teams of leaders from particular schools examine advice- and information-seeking networks in particular schools for particular school subjects. By comparing and contrasting networks between schools or between subjects within schools, participants generate hypotheses about what might be going on in the school and suggest additional diagnostic questions they would like to pursue. Using simulations, we press practitioners to engage in design work grounded in their diagnosis of particular situations (e.g., "Imagine that you will introduce a new mathematics curriculum into these two schools, using this network data from the schools. How might the data affect your implementation strategy, and why?").

Moving beyond engaging practitioners with cases of other schools, we use our instruments to transfer knowledge and skills from anonymous school cases to participants' own schools. For

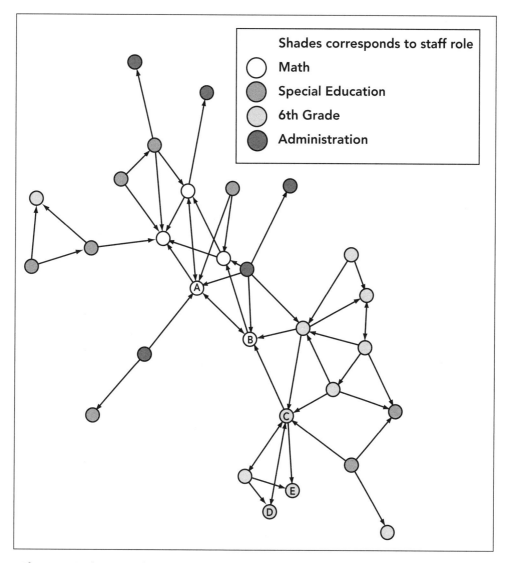

Fig. 1.4: Sociogram of a middle school mathematics advice-and-information network

example, after using social network diagrams from case-study schools to diagnose interaction patterns in essential school subjects and design improvement efforts, we have participants complete the social network instrument for their own school. They then map their social network data and compare their map to those from other schools, identifying commonalities and differences and developing hypotheses about them. For example, most of the organizational routines in our school do not involve teachers from different grades. That pattern may account for why very little communication about mathematics occurs across grade levels.

We have worked to engage study participants with data from their own schools. We accomplish this by composing individualized school reports based on our data and then by conducting structured workshops that have school leaders use these reports in diagnostic work. Figure 1.4 is an example of the sort of network data maps that we generate. Examining figure 1.4, a school

leader might notice that node A is crucial in linking mathematics teachers with special education teachers, whereas the paired relationship between nodes B and C are a crucial link, though
not the only one, in mathematics advice-and-knowledge relations between middle grade mathematics teachers and generalist sixth-grade teachers. Indeed, practitioners might wonder why
interactions among staff about mathematics education in this middle school are chiefly, though
not exclusively, contained either among middle school teachers or among sixth-grade teachers,
with fewer interactions between these two categories.

Possibilities and Problems in Partnering

The three examples of partnering discussed in this chapter capture some of the ways in which
the DLS have forged connections between research and development efforts on the one hand
and policy and practice on the other. Many similarities cross the three cases, but so do important
differences. In some instances, as in our work with CPS, we have forged direct ties to district
policymakers and school practitioners. In others, such as with M[2] and the Penn Center, we
have forged ties with practitioners and district policymakers indirectly through a third party—a
university-based research or development project, or both, working in collaboration with a local
school district. For example, in our partnership with M[2], connections between DLS and school
practitioners were indirect, enabled by our colleagues at UNL. From the outset, beginning
with tailoring SSSNQ questions and items, our colleagues at UNL guided our efforts in both
research and development. By partnering with this in-state and locally known research-and-development entity, the DLS gained access to and legitimacy with local policymakers and practitioners. We believe that these partnering efforts improved both the relevance of our instruments
to state and local conditions and the out-of-state response rates to the SSSNQ.

Our partnering efforts went beyond a classic collaboration across institutions or disciplinary traditions. They thus encountered some partnership difficulties concerning connecting with
developers, policymakers, and practitioners. For example, although our research designs chiefly
aimed to generate hypotheses rather than test them, many district policymakers tried to draw
causal inferences and look for any significant school-level variables that might account for particular outcomes. In such situations, one must constantly remind people that you cannot draw
causal inferences from these data. Further, encouraging people to think about what sort of data
they would need to support a causal claim about a particular relationship not only reminds
people of the limits of particular data but also engages them in thinking about different research
designs. As researchers, we faced the challenge of maintaining the boundaries of our study designs and research findings while also fostering good relations with our partners. In this respect,
the efforts of the DLS to connect with policymakers and practitioners were not unusual.

However, from our efforts to use our research data to engage practitioners and policymakers
in developing practical knowledge through diagnosis and design using a distributed framework,
a somewhat unique obstacle emerged. Writing research reports based on empirical data from
a particular school, we faced the challenge of balancing our desire to provide relevant data to
school leaders with the imperative to protect study participants' confidentiality. Such confidentiality must be protected, not only for compliance with the requirements of Human Subjects

Research Boards, but also to maintain a trusting relationship with study participants. If participants believe that sharing findings with their schools has breached the promise of confidentiality, they probably will not participate in future rounds of data collection, or they may participate inauthentically. Our instance heightened these challenges because the social network items on the SSSNQ asked school staff to name those from whom they seek advice and information about core aspects of their classroom work. Protecting the confidentiality of study participants in this situation is neither simple nor straightforward. For example, if a teacher identifies a colleague as a source of advice, but that colleague has not consented to participate in the research—for example, if he or he did not respond to the survey—can that analysis consider that relationship without violating the advice source's confidentiality?

Sharing social network data from a school with the leaders from that school, so the leaders engage in relevant diagnostic and design work, made such confidentiality problems even more pronounced. In order to share findings from our analysis of the social network data without violating confidentiality, we constructed categories of school staff that included enough staff members to make it impossible to determine the identity of any one staff member. For example, see figure 1.4, which captures the sociogram depictions for one school where the DLS have shared back data with study participants. Here, circles representing teachers are colored according to the teacher's role, so that any one category stands for at least five people. Similarly, our quantitative analysis of the network data reported averages across categories containing at least five people. In our experience, analyzing social network data is valuable for engaging school staff in diagnostic work about their school's leadership and management, but it also raises concerns about confidentiality (Borgatti and Molina 2003).

Presenting these data to school practitioners, we observed that, when analyzing a sociogram representation of the social network data from their school, the impulse was to try to assign names to each of the nodes (e.g., node B or C in fig. 1.4). Although such attempts to breach confidentiality always troubled us, we believed the attempts to be speculative at best, because even if the data provoked guessing games, the data sufficiently concealed the identities of individual participants. We counseled study participants accordingly. Also, to discourage misinterpreting the data, we emphasized during our share-back presentation that the social-network data, like all survey measures, contained some measurement error. One should thus interpret the social network data only as one potentially limited representation of interactions among school staff.

Conclusion

The DLS sought to understand leadership and management in schools as well as engage practitioners and policymakers in improving the practice of leading and managing. We employed numerous research methods to develop hypotheses about leading and managing instruction—research findings that centered on *what* knowledge. We also ventured beyond generating conventional research findings for consumption by fellow researchers and focused on reaching practitioners and policymakers to develop *how* knowledge. Our research and development, enabled by partnering with practitioners, policymakers, and researchers, has centered on designing tools that develop practical knowledge about leading and managing instruction in mathematics and other school subjects. We hope that this work on the knowledge of practice has guided, and will continue to guide, policy and practice. The work has likewise influenced our research, pressing us to tailor our methods in order to design learning opportunities for practitioners and policymakers.

REFERENCES

Borgatti, Stephen P., and Jose Luis Molina. "Ethical and Strategic Issues in Organizational Network Analysis." *Journal of Applied Behavioral Science* 39 (September 2003): 337–49.

Burch, Patricia, and James P. Spillane. "Elementary School Leadership Strategies and Subject Matter: Reforming Mathematics and Literacy Instruction." *Elementary School Journal* 103 (May 2003): 519–35.

Burt, Ronald S. "The Network Structure of Social Capital." In *Research in Organizational Behavior*, edited by Barry M. Staw and Robert I. Sutton, pp. 354–423. New York: Elsevier Science, Inc., 2000.

Camburn, Eric, James P. Spillane, and James Sebastian. "Investigating the Validity of a Daily Log and Its Utility for Assessing the Impact of Programs on Principals." *American Journal of Education,* under review.

Coburn, Cynthia E., and Jennifer L. Russell. "District Policy and Teachers' Social Networks." *Educational Evaluation and Policy Analysis* 30 (September 2008): 203–35.

Collins, Allan, Diana Joseph, and Katerine Bielaczyc. "Design Research: Theoretical and Methodological Issues." *Journal of the Learning Sciences* 13 (January 2004): 15–42.

Drake, Corey, James P. Spillane, and Kimberley Hufferd-Ackles. "Storied Identities: Teacher Learning and Subject-Matter Context." *Journal of Curriculum Studies* 33 (January 2001): 1–23.

Firestone, William A. "Using Reform: Conceptualizing District Initiative." *Educational Evaluation and Policy Analysis* 11 (January 1989): 151–64.

Grossman, Pamela L., and Susan S. Stodolsky. "Content as Context: The Role of School Subjects in Secondary School Teaching." *Educational Researcher* 24 (November 1995): 5–11, 23.

Pitts, Virginia, and James P. Spillane. "Using Social Network Methods to Study School Leadership." *International Journal of Research and Method in Education* 32 (July 2009): 185–207.

Price, Jeremy N., and Deborah Loewenberg Ball. "'There's Always Another Agenda': Marshalling Resources for Mathematics Reform." *Journal of Curriculum Studies* 29 (November 1997): 637–66.

Pustejovsky, James, and James P. Spillane. "Question-Order Effects in Social Network Name Generators." *Social Networks* 31 (October 2009): 221–29.

Pustejovsky, James, James P. Spillane, Ruth M. Heaton, and William J. Lewis. "Understanding Teacher Leadership in Middle School Mathematics: A Collaborative Research Effort." *Journal of Mathematics and Science: Collaborative Explorations* 11 (Spring 2009): 19–40.

Reagans, Ray, and William McEvily. "Network Structure and Knowledge Transfer: The Effects of Cohesion and Range." *Administrative Science Quarterly* 48, no. 2 (2003): 240–67.

Rowan, Brian, Stephen W. Raudenbush, and Yuk Fai Cheong. "Teaching as a Nonroutine Task: Implications for the Organizational Design of Schools." *Education Administrations Quarterly* 24 (November 1993): 479–500.

Spillane, James P. *Distributed Leadership*. San Francisco: Jossey-Bass, 2006.

———. "Primary School Leadership Practice: How the Subject Matters." *School Leadership and Management* 25 (September 2005): 383–97.

Spillane, James P., Eric M. Camburn, and Amber Stitziel Pareja. "Taking a Distributed Perspective to the School Principal's Workday." *Leadership and Policy in Schools* 6 (February 2007): 103–25.

Spillane, James P., and Amy F. Coldren. *Managing to Lead: Taking a Distributed Perspective to Diagnosis and Design in School Leadership and Management*. New York: Teachers College Press, in progress.

Spillane, James P., and Anita Zuberi. "Designing and Piloting a Leadership Daily Practice Log: Using Logs to Study the Practice of Leadership." *Educational Administration Quarterly* 45 (August 2009): 375–23.

Spillane, James P., and John B. Diamond. *Distributed Leadership in Practice*. New York: Teachers College Press, 2007.

Spillane, James P., Bijou Hunt, and Kaleen Healey. "Managing and Leading Elementary Schools: Attending to the Formal and Informal Organization." *International Studies in Educational Administration* 37 (2009): 5–28.

Spillane, James P., Richard Halverson, and John B. Diamond. "Investigating School Leadership Practice: A Distributed Perspective." *Educational Researcher* 30 (April 2001): 23–28.

———. "Toward a Theory of Leadership Practice: A Distributed Perspective." *Journal of Curriculum Studies* 36, no. 1 (2009): 3–34.

Spillane, James P., Tim Hallett, and John B. Diamond. "Forms of Capital and the Construction of Leadership: Instructional Leadership in Urban Elementary Schools." *Sociology of Education* 76 (January 2003): 1–17.

Stodlosky, Susan S. *The Subject Matters: Classroom Activity in Math and Social Studies*. Troy, N.Y.: Educator's International Press, 1998.

Surveys of Enacted Curriculum and the Council of Chief State School Officers Collaborative

Andrew C. Porter
Jennifer McMaken
Rolf K. Blank

T HE UNITED STATES has a long history of interest in the content of grades K–12 math instruction. Beginning with the curricular reforms in the 1970s, researchers have sought to strengthen the quality of mathematics instruction. The release of the National Council of Teachers of Mathematics's (NCTM) *Curriculum and Evaluation Standards for School Mathematics* (NCTM 1989) significantly strengthened these efforts. *Curriculum and Evaluation Standards* gave a clear overview of what mathematical literacy entails and codified a set of standards for instruction content that could help guide curricular reform.

With the No Child Left Behind Act up for reauthorization, strong indications exist that standards-based reform will continue along with the discussion of what grades K–12 mathematics teaches. Organizations such as NCTM have compiled updated standards for mathematics instruction based on input from expert panels for grades K–8 (NCTM 2006) and, more recently, for high school (NCTM 2009). Similarly, a partnership between the National Governor's Association and the Council of Chief State School Officers (CCSSO) (2010) is currently developing a set of common standards for language arts and mathematics that has the support and participation of 35 states.

A central challenge in using standards is determining how to measure alignment among content standards, assessments, and instruction. A few approaches to measuring alignment exist. Most start with a particular state's standards and ask how much of a standard's content a given curriculum or assessment contains. Most of these approaches, however, do not extend to investigating instruction (Webb 1997, 2002). An alternative approach, developed from the work of the Content Determinants group at the Institute for Research on Teaching at Michigan State University (Porter et al. 1978) in partnership with the Wisconsin Center for Education Research (WCER) and the CCSSO, is the Surveys of Enacted Curriculum (SEC).

This chapter presents a case study of the State Collaborative on Assessment and Student

Standards' (SCASS) use of the SEC. This collaborative, led by the CCSSO, sought to enhance states' and districts' capability for analyzing the alignment of instructional practices and content taught in classrooms to required standards and assessments of students' learning. This partnership's work derived from the availability of a tool that provides reliable information about alignment. We therefore begin with an overview of what the SEC are, how they can be used, and of what quality they are. The remainder of the chapter then discusses how the SCASS used the SEC and other resources to improve educators' ability to adapt and strengthen their instructional practice. We conclude with a summary of the lessons learned and a set of recommendations for how others may create partnerships between research and practice in ways that strengthen mathematics education.

The Surveys of Enacted Curriculum (SEC)

The SEC are a practical, reliable set of data-collection and reporting tools used in states and schools across the United States. *Enacted curriculum* is the curriculum that a teacher actually teaches in the classroom. The SEC exist for grades K–12 instruction in mathematics, English, language arts, and science, to collect and report data on instructional practices and content. The survey instruments and resulting data allow educators to analyze objectively the degree of alignment, or consistency, among instruction, state content standards, and assessments. Once teachers complete the survey's questions online, the system produces data in user-friendly charts and graphs that facilitate analyzing differences across classrooms, schools, or districts.

State curriculum specialists, teachers, and researchers developed the surveys and reporting tools over the past decade, under CCSSO and WCER leadership. Hundreds of schools and classrooms have field-tested the surveys in collaboration with state education agencies. CCSSO and several partner organizations made the data collection, analysis, and application services available. CCSSO works with states to facilitate content analysis of state standards and assessments as well as of other agencies' documents (e.g., National Assessment of Educational Progress [NAEP] frameworks and assessments). One of these collaborations' central aims is to analyze instruction's alignment with the content called for in standards and assessments.

The SEC Tool

The SEC are a two-dimensional representation of topic and cognitive demand. They divide each subject's content into general topics. Grades K–12 mathematics has sixteen general topics (e.g., operations, measurement, basic algebra) and 217 specific topics arranged across the sixteen general ones (e.g., coordinate planes, inequalities, factoring). The mathematics survey has five levels of cognitive demand: memorize; perform procedures; demonstrate understanding; conjecture, generalize, and prove; and solve nonroutine problems or make connections.

To gather information on instructional content, teachers complete a survey where they indicate, for a given period, (*a*) the amount of time devoted to each topic (coverage), and then, for each topic, (*b*) the relative emphasis given each category of cognitive demand (see fig. 2.1). A four-point scale measures both as follows:

- Coverage: (1) none or not covered, (2) slight (less than one lesson), (3) moderate (one to five lessons), and (4) sustained (more than five lessons);

- Relative emphasis: (1) none, (2) slight (less than 25 percent of the time spent on the topic), (3) moderate (25–33 percent of the time spent), and (4) sustained (more than 33 percent).

After teachers report their instructional content, an algorithm transforms their data into proportions of total instructional time spent on each cell in the content matrix. The proportions, which sum to 1 across the cells of the content matrix, reflect how much relative weight teachers gave specific content over a given time period (Porter and Smithson 2001). Teachers, administrators, and researchers could then compare instructional content across a school's, district's, or state's classrooms, or they could compare the content to a relevant set of standards or assessments.

Time on Topic	Elementary School Mathematics Topics	Expectations for Students in Mathematics				
<none> ⓪ ① ② ③	Number sense / Properties / Relationships	Memorize Facts/ Definitions/ Formulas	Perform Procedures	Demonstrate Understanding of Mathematical Ideas	Conjecture, Generalize, Prove	Solve Non-Routine Problems/Make Connections
⓪ ① ② ③	¹⁰¹ Place value	⓪ ① ② ③	⓪ ① ② ③	⓪ ① ② ③	⓪ ① ② ③	⓪ ① ② ③
⓪ ① ② ③	¹⁰² Patterns	⓪ ① ② ③	⓪ ① ② ③	⓪ ① ② ③	⓪ ① ② ③	⓪ ① ② ③
⓪ ① ② ③	¹⁰³ Decimals	⓪ ① ② ③	⓪ ① ② ③	⓪ ① ② ③	⓪ ① ② ③	⓪ ① ② ③
⓪ ① ② ③	¹⁰⁴ Percent	⓪ ① ② ③	⓪ ① ② ③	⓪ ① ② ③	⓪ ① ② ③	⓪ ① ② ③
⓪ ① ② ③	¹⁰⁵ Real numbers	⓪ ① ② ③	⓪ ① ② ③	⓪ ① ② ③	⓪ ① ② ③	⓪ ① ② ③
⓪ ① ② ③	¹⁰⁶ Exponents, scientific notation	⓪ ① ② ③	⓪ ① ② ③	⓪ ① ② ③	⓪ ① ② ③	⓪ ① ② ③
⓪ ① ② ③	¹⁰⁷ Factors, multiples, divisibility	⓪ ① ② ③	⓪ ① ② ③	⓪ ① ② ③	⓪ ① ② ③	⓪ ① ② ③
⓪ ① ② ③	¹⁰⁸ Odds, evens, primes, composites	⓪ ① ② ③	⓪ ① ② ③	⓪ ① ② ③	⓪ ① ② ③	⓪ ① ② ③
⓪ ① ② ③	¹⁰⁹ Estimation	⓪ ① ② ③	⓪ ① ② ③	⓪ ① ② ③	⓪ ① ② ③	⓪ ① ② ③
⓪ ① ② ③	¹¹⁰ Order of operations	⓪ ① ② ③	⓪ ① ② ③	⓪ ① ② ③	⓪ ① ② ③	⓪ ① ② ③
⓪ ① ② ③	¹¹¹ Relationships between operations	⓪ ① ② ③	⓪ ① ② ③	⓪ ① ② ③	⓪ ① ② ③	⓪ ① ② ③

Fig. 2.1. SEC content matrix on teacher survey

The SEC's two-dimensional language can also analyze the content of instructional materials, standards, and assessments. A primary decision is which unit to analyze. For instruction, the analysis unit is time on topic. For assessments, the unit is each item. For content standards, selecting the unit is more difficult. The most successful approach has been to pick the most specific version of the content standards and, within that, analyze the content of each objective, paragraph, or phrase (Porter 2002). Trained content analysts rate each item on a test or each objective in the standards on what content it represents. Three to five content analysts work independently to code the items, then convene to discuss any noted problems or questions. After that, they give their final ratings for all items under review.

The SEC are useful to a variety of audiences, including school administrators, teachers, test and curriculum developers, and researchers. Perhaps the SEC's most compelling use is in increasing the knowledge of how instructional content relates to standards. CCSSO/SCASS has used the SEC in work with participating states to investigate their content standards, students' achievement tests, and instructional practices (Porter, Polikoff, and Smithson 2009). To date, more than thirty states have participated in SCASS's use of the SEC. This unique collaboration has created a means of exploring how research and policy intersect to influence practice in mathematics education in states around the country.

Evidence Quality

Thus far we have discussed how we can use the SEC to provide powerful descriptions of content emphases and the degree of content overlap among instruction, interventions, assessments, and content standards. We have said little, however, about the validity of SEC data. The typical response rate for the SEC teachers' survey is 75 percent, even for national probability samples of teachers (i.e., large samples of teachers selected from across the country so that different areas are represented in proportion to their makeup within the population) (Garet et al. 2001). More impressive, the response rate stayed at 75 percent in a longitudinal study that required teachers to complete the survey once yearly for three consecutive years (Porter et al. 2000). Also, a number of studies have examined the validity of teachers' reports of content in mathematics instruction from sources other than the SEC. These studies have generally found such reports to be quite accurate describing instruction's quantity (the focus of SEC), but not its quality (Burstein et al. 1995; Desimone 2009; Herman, Klein, and Abedi 2000; Mayer 1999; McCaffrey et al. 2001; Spillane and Zeulix 1999). We have investigated this finding by using the SEC with a sample of 62 teachers in twelve districts across six states where we found high rates of agreement between observations and teachers' reports (correlations of 0.6–0.8) (Porter et al. 1993).

Reliabilities among ratings for trained content analysts were also high. In one study, the reliability of ratings was 0.70 across two raters and 0.82 across four raters (Porter 2002; Porter, et al. 2008). Surprisingly, agreement among raters for assessments was no higher than for content standards. Increasing the number of raters beyond four did not substantially increase reliability.

Another source of validity for the SEC comes from using the instrument to predict gains in students' achievement. If teachers' SEC reports of what they teach strongly predict gains in students' achievement, then the SEC reports must be valid. In a study of upgrading high school mathematics, Gamoran et al. (1997) found that the degree of alignment between teachers' SEC reports of the content taught and the measure of students' achievement strongly predicted differential gain in students' achievement across teachers in the study (correlation 0.45). Gamoran et al. found this correlation to be true only when defining content at the intersection of topic and cognitive demand; when mathematics content was described only by topics or only by cognitive demand, the correlation dropped to near zero.

Yet another study found that data on curriculum content reported in teachers' surveys covering a whole year correlated closely with data aggregated from daily logs of instructional content (Smithson and Porter 1994).

The SEC State Collaborative on Assessment and Student Standards (SCASS)

The partners in the SEC SCASS are staff of the CCSSO, who support the work of the collaborative; state education agencies that pay membership dues for services from the SEC state collaborative; the WCER, who manage the content analysis process by scoring and analyzing the SEC data; and service-provider organizations and consultants. Several research and education services organizations work with the SEC collaborative to provide professional development, leadership training, and technical assistance on using SEC tools and analyzing the data that the tools produce.

State education agencies initially conceived of the survey forms and items as a method of measuring opportunity to learn in relation to state assessments in science and math. Grants from the National Science Foundation (NSF) and state education agencies' voluntary annual memberships funded the surveys' initial development and testing. Porter and Smithson's *Research Up Close* study (1994), supported by an NSF grant, developed and tested a framework for analyzing instruction content. Educators and researchers participating in the SEC collaborative in 1999 reviewed and revised the framework (Blank, Porter, and Smithson 2001). Later, funded by a math-science partnership RETA grant, the collaborative revised and upgraded survey items on teachers' professional development and instructional practices (Smithson and Blank 2007).

Goals of the SCASS

One goal of the SCASS is to assist states and local districts by giving them a practical, reliable set of data-collection and reporting tools, ones that can help educators analyze the alignment of instructional practices and content taught in classrooms with required standards and assessments of students' learning. A second goal is to provide professional development and leadership to state education managers and local educators on the uses of enacted curriculum data to make decisions about how to improve instruction. A third is to furnish data that helps evaluate change meaningfully over time, especially related to an education intervention or an improvement initiative. Developers designed the survey system to provide longitudinal data at the class, school, and district level that teachers and administrators can use to evaluate instructional changes related to specific initiatives.

Supporting Arrangements

The SEC have grown from a research project in the mid-1990s to a fully operational collaborative that renders data and alignment services in mathematics, English language arts, and science to more than thirty states. In the past year, the SEC collected and reported data from more than 10,000 teachers, 4,000 of whom taught mathematics. The SEC collaborative continues to offer member agencies and educators crucial data for supporting education initiatives and measuring the initiatives' impact. Professional development and training in the use of SEC data now focuses on linking the analysis and use of instructional data to school improvement plans and initiatives.

Central Policy Levers

SCASS has worked at state policy and local levels. States define and establish their content standards and assessments. State education agencies define and implement program initiatives in curriculum, professional development, school improvement, assessment, and accountability. The SEC collaborative operates in each of these areas to compile data related to classroom instruction's alignment with state standards and assessments. State agencies can request and support content analysis of standards and assessments through the SEC in order to conduct instructional alignment studies. Local districts can contract directly with the SEC to access and use the surveys to (1) collect and analyze data on instruction in core subject areas, (2) focus improvement efforts, and (3) analyze the degree of change and improvement in alignment to standards.

A Theory of Policy Influence

Porter and colleagues (Porter et al. 1988; Porter 1994) presented one way to conceptualize the process through which the SEC SCASS may influence policy. They posited a theory of five attributes that policies must have in order to influence teachers' decisions about what to teach. The theory hypothesized that, in the instance of the SCASS and the SEC, standards-based reform would have greatest influence on instructional content (i.e., what is taught). This influence would exist to the extent that the enacted polices (1) were specific about what to teach, (2) were consistent in the message they sent, (3) had authority, (4) had power, and (5) were stable over time.

The first two points are transparent. The authority level ascribed to policies derives from the extent to which experts develop and promote the policies, states adopt them, and current teaching beliefs and practice follow them. Policies have power to the extent that they reward compliance and penalize noncompliance. Researchers have applied this theory of policy influence on teachers' practice to reviews of systemic reform and to comprehensive school reform (Berends, Bodilly, and Kirby 2002; Clune et al. 1993; Desimone 2002). The SEC tools help create policies (e.g., content standards and assessment) that are appropriately *specific* and *consistent*. The tools are also useful for monitoring teachers' instructional practices to see if, over time, the practices align increasingly to content standards. As instruction aligns more to content standards, students' achievement on related tests improves.

An SEC-Based Intervention to Guide Practice and Its Evaluation

The SEC collaborative developed a model for data-driven improvement of instruction focused on analyzing instruction's alignment to required standards for teaching. Initially, a three-year, experimentally designed study using randomized, in-school trials tested the SEC-based instructional improvement model. The improvement model included professional development in which school leaders and teachers analyzed, interpreted, and reflected on their own instruction by using data from SEC reports as well as achievement data from state or district assessments. The data analyses guided decisions to refocus instruction through improvements in teaching, curriculum materials, and conditions for teaching. Fifty U.S. middle schools in five large, urban districts participated in the study, with half the schools in each district randomly assigned to receive the two-year treatment and half assigned to a control group (Porter et al. 2007). Each treatment school formed an improvement leadership team of five to seven members, among them teachers, subject specialists, and at least one administrator. Teams received professional development on data analysis and instructional leadership. The teams then trained and gave technical assistance to all math and science teachers in their schools (Blank et al. 2006). Teachers used the SEC data to analyze their own instruction, relating their instruction to state content standards and assessment and comparing it to that of other teachers in their school or district. Over time, they used the SEC to evaluate change in instruction. Data analysis showed a statistically significant (0.36 effect size) increase in mathematics instruction's alignment to mathematics state content standards (Porter et al. 2007).

Features and Outcomes of the SCASS

SCASS has developed, and maintained its currency, through a close collaboration among researchers, CCSSO staff, and WCER staff. Members of these entities frequently partner on work that further extends SCASS's lessons and findings to a broader audience. Having WCER conduct all the content analyses had the advantage of creating an available, national data repository of SEC content analysis results. The national repository allowed partners to examine easily how math standards in one state compare to those in another or to those of other targets such as NCTM or NAEP. For example, recent work using the national SEC data system allowed an examination of (1) the extent to which a de facto national curriculum exists among state content standards (Porter, Polikoff, and Smithson 2009) and (2) the extent to which a de facto national, assessed curriculum exists (Porter and Polikoff, under review).

Research findings presented to SCASS participants, from the field as well as from the collaborative, guide future efforts to improve instruction. States in the collaborative receive ongoing mentorship on both individual state progress and new advances in the field, thus continuing the cycle whereby practice and research influence each other. By its very nature, the SCASS convenes people who greatly influence how a state enacts educational policy. By primarily engaging state educational leaders, the SCASS positions itself well to influence how research affects policy enactment.

The process for creating sound standards and assessments is neither simple nor fast. High-quality standards require a lengthy development process to ensure the appropriate levels of comprehensiveness and rigor. Similarly, creating assessments is expensive and time consuming. Moreover, research has shown that many states lack expertise in assessing the alignment of their standards and tests. Porter, Polikoff, and Smithson (2009) have shown that standards and tests do not automatically align strongly. They noted that, although some states explicitly claimed that their content standards development drew on NCTM's standards, these state's standards aligned no better to NCTM's than those of states that made no such claim. Since these national professional standards, developed with input from a broad array of experts, are generally held to be of high quality, such a pervasive low level of alignment with them is discouraging. The limitations of human judgment from which that low alignment level derives point to both the benefits of objective measures such as the SEC and the need to think more carefully about influences on policy. One way to strengthen the alignment of state standards to NCTM standards could be to engage more directly with policymakers and state education leaders.

Recommendations for the Field

Experience from the SEC collaborative leads us to see a valuable role for cross-institution partnerships. Researchers continue to develop new tools and instruments to conduct their research. Developing tools with value for improving policy and practice, however, often requires that researchers partner with organizations equipped and skilled in development and outreach. The developers of the SEC first partnered with CCSSO. Later, we partnered with Learning Points Associates (LPA) to extend the content matrix to English, language arts, and reading. In other instances, researchers may partner with a publisher to distribute a tool. Once a tool advances to replication and dissemination, the researcher should give thought to what outcomes he wishes to achieve and what method of reproduction will best achieve those outcomes. For example, we decided to keep the SEC's tools in public domain.

One important feature of the SCASS is the way it supports improving mathematics instruction: it nurtures a professional community of state education leaders from around the country. Participating states come together not only to analyze the content of education in their states, but also to share best practices on approaches taken to meet a need. The SCASS collaborative convenes biannually for project leadership planning and training meetings for state and district leaders who are members. The collaborative also works with state leaders who are members to plan and organize training workshops in states and districts on using the SEC and the data it reports at the local level. Finally, through the SCASS, CCSSO works with states to develop multistate proposals to federal agencies and foundations for funding research and development projects. These projects seek to improve the system of data collection and reporting further and to deliver important technical assistance and training for educators and leaders on data-driven instructional improvement. (For further information on the SEC collaborative from CCSSO, see www.SECsurvey.org.)

Connecting research, practice, and policy—the focus of this volume—is the holy grail for education: it happens, but not easily or as often as would be optimal. We have described how states and others have used SEC tools to investigate and improve the properties of their standards-based reform policies and to provide guidance and professional development to teachers.

We initially developed the SEC tools as measures of how important variables in research affect teachers' decisions about what to teach in mathematics. Why, then, did the SEC tools become so visible and popular among practitioners and policymakers? One answer is that the tools have potential for improving practice, but that alone would not explain their relatively widespread use and influence. Many tools developed in research go no further. Another answer, found in a recent article on making connections between research and practice (Porter and McMaken 2009), offers six hypotheses, the last of which is "Formal and informal networks exist among leaders in education practice. Largely through word of mouth, reform ideas pass quickly from one leader to another in these networks and on into implementation" (p. 63). Explaining the SEC tools' popularity points to several interconnected networks. The tools have their origin in Andrew Porter's research on how teachers make decisions about what to teach in mathematics. Rolf Blank, a boundary crosser perfectly positioned at the CCSSO, is a leader of the SCASS networks. At one point, he had a project on indicators in which he had involved Porter as a consultant. Together and with others, Porter and Blank collaborated on a number of projects with NSF support, with Blank extending the indicators' work into SCASS work. Blank has since then been the center of the SEC project's activity. Through the SCASS network, he tapped the networks of state leaders in assessment and evaluation across the country. John Smithson, who maintains the SEC database and analysis capability, was a graduate assistant under Porter at the University of Wisconsin—Madison. Later, LPA picked up on the SEC tools as powerful mechanisms for use in teachers' professional development. Since the SEC were only available in mathematics and science and LPA needed a tool for English, language arts, and reading, LPA extended the SEC tools to meet their need. SEC's success has come from bringing together formal and informal networks based on common interests, collaborative work, and collegial relationships.

Researchers typically are not positioned well for connecting research and practice by themselves. Understandably, their goals are to do research on important problems in education and publish the results. Boundary crossers at professional organizations (e.g., CCSSO), regional labs (e.g., LPA), and other organizations, however, can and have built networks of researchers and practitioners. As we have stated in describing the SEC study, boundary crossers do the development work that puts research findings and tools into the hands of capable practitioners. Such work has not only improved practice but also guided research's interpretation and future directions.

REFERENCES

Berends, Mark, Susan J. Bodilly, and Sheila Nataraj Kirby. *Facing the Challenges of Whole-School Reform: New American Schools after a Decade.* Santa Monica, Calif.: RAND Corp., 2002.

Blank, Rolf K., Andrew C. Porter, and John L. Smithson. *New Tools for Analyzing Teaching, Curriculum, and Standards in Mathematics and Science: Results from Survey of Enacted Curriculum Project.* Washington, D.C.: Council of Chief State School Officers, 2001.

Blank, Rolf K., John L. Smithson, Andrew C. Porter, Diana Nunnaley, and Eric Osthoff. "Improving Instruction through Schoolwide Professional Development: Effects of the Data-on-Enacted-Curriculum Model." *ERS Spectrum Journal of Research and Information* 24 (Spring 2006): 9–23.

Burstein, Leigh, Lorriane M. McDonnell, Jeanette Van Winkle, Tor H. Ormseth, Jim Mirocha, and Gretchen Guiton. *Validating National Curriculum Indicators.* Santa Monica, Calif.: RAND Corp., 1995.

Clune, William H. "Systemic Educational Policy: A Conceptual Framework." In *Designing Coherent Education Policy: Improving the System,* edited by Susan H. Fuhrman, pp. 125–40. New York: Jossey-Bass, 1993.

Desimone, Laura. "How Can Comprehensive School Reform Models Be Successfully Implemented?" *Review of Educational Research* 72 (Fall 2002): 433–79.

Gamoran, Adam, Andrew C. Porter, John L. Smithson, and Paula A. White. "Upgrading High School Mathematics Instruction: Improving Learning Opportunities for Low-Achieving, Low-Income Youth." *Educational Evaluation and Policy Analysis* 19 (Winter 1997): 325–38.

Garet, Michael S., Andrew C. Porter, Laura Desimone, Beatrice F. Birman, and Kwang Suk Yoon. "What Makes Professional Development Effective? Results from a National Sample of Teachers." *American Educational Research Journal* 38 (Winter 2001): 915–45.

Herman, Joan L., Davina C. D. Klein, and Jamal Abedi. "Assessing Students' Opportunity to Learn: Teacher and Student Perspectives." *Educational Measurement: Issues and Practice* 19 (Winter 2000): 16–24.

Mayer, Daniel P. "Measuring Instructional Practice: Can Policymakers Trust Survey Data?" *Educational Evaluation and Policy Analysis* 21 (Spring 1999): 29–45.

McCaffrey, Daniel F., Laura S. Hamilton, Brian M. Stecher, Stephen P. Klein, Delia Bugliari, and Abby Robyn. "Interactions among Instructional Practices, Curriculum, and Student Achievement: The Case of Standards-Based High School Mathematics." *Journal for Research in Mathematics Education* 22 (November 2001): 493–517.

National Council of Teachers of Mathematics (NCTM). *Curriculum and Evaluation Standards for School Mathematics.* Reston, Va.: NCTM, 1989.

———. *Curriculum Focal Points for Prekindergarten through Grade 8 Mathematics: A Quest for Coherence.* Reston, Va.: NCTM, 2006.

———. *Focus in High School Mathematics: Reasoning and Sense Making.* Reston, Va.: NCTM, 2009.

National Governor's Association and the Council of Chief State School Officers (NGA/CCSSO). "College- and Career-Readiness Standards." Washington, D.C.: Common Core State Standards Initiative, 2010. http://www.corestandards.org/the-standards/.

———. "National Standards and School Improvement in the 1990s: Issues and Promise." *American Journal of Education* 102 (August 1994): 421–49.

———. "Curriculum Reform and Measuring What Is Taught: Measuring the Quality of Education Processes." Paper presented at the annual meeting of the Association for Public Policy Analysis and Management, New York, October 1998.

———. "Measuring the Content of Instruction: Uses in Research and Practice." *Educational Researcher* 31 (October 2002): 3–14.

Porter, Andrew C., and Jennifer McMaken. "Making Connections between Research and Practice." *Phi Delta Kappan* 91 (September 2009): 61–64.

Porter, Andrew C., and John L. Smithson. "Are Content Standards Being Implemented in the Classroom? A Methodology and Some Tentative Answers." In *From the Capitol to the Classroom: Standards-Based Reform in the States.* One Hundredth Yearbook of the National Society for the Study of Education, Part 2, edited by Susan H. Fuhrman, pp. 60–80. Chicago: University of Chicago Press, 2001.

Porter, Andrew C., and Morgan S. Polikoff. "The Role of State Student Achievement Tests in Standards-Based Reform." Under review.

Porter, Andrew C., John L. Smithson, Rolf K. Blank, and Timothy Zeidner. "Alignment as a Teacher Variable." *Applied Measurement in Education* 20, no. 1 (2007): 27–51.

Porter, Andrew C., Michael S. Garet, Laura Desimone, Kwang Suk Yoon, and Beatrice F. Birman. *Does Professional Development Change Teaching Practice? Results from a Three-Year Study.* Report to the U.S. Department of Education, Office of the Under Secretary, on Contract No. EA97001001 to the American Institutes for Research. Washington, D.C.: Pelavin Research Center, 2000.

Porter, Andrew C., Michael W. Kirst, Eric J. Osthoff, John L. Smithson, and Steven A. Schneider. *Reform Up Close: An Analysis of High School Mathematics and Science Classrooms.* Final report to the National Science Foundation on Grant No. SPA-8953446 to the Consortium for Policy Research in Education. Madison: University of Wisconsin—Madison, Wisconsin Center for Education Research, 1993.

Porter, Andrew C., Morgan S. Polikoff, and John L. Smithson. "Is There a De Facto National Intended Curriculum? Evidence from State Content Standards." *Educational Evaluation and Policy Analysis* 31 (September 2009): 238–68.

Porter, Andrew C., Morgan S. Polikoff, Timothy Zeidner, and John L. Smithson. "The Quality of Content Analyses of State Student Achievement Tests and State Content Standards." *Educational Measurement: Issues and Practice* 27 (November 2008): 2–14.

Porter, Andrew C., Robert E. Floden, Donald J. Freeman, William H. Schmidt, and John R. Schwille. "Content Determinants in Elementary School Mathematics." In *Perspectives on Research on Effective Mathematical Teaching*, edited by Douglas A. Grouws and Thomas J. Cooney, pp. 96–113. Hillsdale, N.J.: Lawrence Erlbaum Associates, 1988.

Porter, Andrew C., William H. Schmidt, Robert E. Floden, and Donald J. Freeman. "Practical Significance in Program Evaluation." *American Educational Research Journal* 15 (Fall 1978): 529–39.

Smithson, John L., and Andrew C. Porter. *Measuring Classroom Practice: Lessons Learned from Efforts to Describe the Enacted Curriculum: The Reform-Up-Close Study.* Consortium for Policy Research in Education (CPRE) Series Report #31. New Brunswick, N.J.: Rutgers University, CPRE, 1994.

Smithson, John L., and Rolf K. Blank. *Indicators of Quality of Teacher Professional Development and Instructional Change Using Data from Surveys of Enacted Curriculum: Findings from the NSF MSP-RETA Project.* Washington, D.C.: Council of Chief State School Officers, 2007.

Spillane, James P., and John S. Zeuli. "Reform and Teaching: Exploring Patterns of Practice in the Context of National and State Mathematics Reforms." *Educational Evaluation and Policy Analysis* 21 (Spring 1999): 1–27.

Webb, Norman L. *Criteria for Alignment of Expectations and Assessments in Mathematics and Science Education.* Research Monograph No. 6. Madison: University of Wisconsin—Madison, National Institute for Science Education, 1997.

———. *Alignment Study in Language Arts, Mathematics, Science, and Social Studies of State Standards and Assessments for Four States.* Washington, D.C.: Council of Chief State School Officers, 2002.

The Role of Tools in Bridging Research and Practice in an Instructional Improvement Effort

Mary Kay Stein
Jennifer Russell
Margaret Schwann Smith

D EBATES regarding the relationship between research and practice have a long history (Atkinson and Jackson 1997; Brown et al. 1999; Burkhart and Schoenfeld 2003; Kaestle 1993; Kennedy 1997). Mathematics education has not used research to improve practice for a broad and complex set of reasons: the nature of research; the local, situated nature of practice; the training and expected roles of researchers and practitioners; and our limited, unimaginative ways of thinking about how research and practice might be configured with each other (Coburn and Stein 2010).

Here, we focus on this final challenge. We aim to expand mathematics educators' ideas of the pathways and mechanisms that can link research and practice. Specifically, we focus on tools and the routines surrounding their use as a way of bridging research and practice to improve teaching and learning. Tools are artifacts (e.g., curriculum materials, computer-based programs, observation protocols, rubrics) that embody research knowledge in ways in which practice can use them directly. Because tools are positioned—in theory—to influence large numbers of teachers and classrooms (Ball and Cohen 1996), they can be important carriers of research knowledge in large-scale efforts to improve practice.

Despite their promise as vehicles for facilitating large-scale, research-based improvement, how *teachers learn* to improve their practice using tools and the role of tools in school-level, *organizational change processes* remain underspecified. Typically, research and development efforts aim to enhance students' learning by producing tools for students to use in the classroom (e.g., a new

Work on this paper was supported by the Collaborative, Technology-Enhanced Lesson Planning as an Organizational Routine for Continuous, School-Wide Instructional Improvement project, which is funded by a research grant from the Institute for Education Sciences (award number R305A090252). All opinions and conclusions expressed in this paper are those of the authors and do not necessarily reflect the views of any funding agency.

curriculum, a computer program). Developers view teachers as conduits through which tools reach students, not as learners themselves who are situated in a complex work environment. Even when teachers' learning is a tool's primary purpose (e.g., with tools such as reflection templates and cases), developers typically devote more attention to how teachers *should use the tool* than to how teachers might *learn to improve practice by using the tool* (Fishman and Davis 2006). This flies in the face of what is known about the extensive learning that teachers require for effective implementation of new practices (Cohen and Barnes 1993). The learning demands on teachers are especially intense when the kind of improvement being sought is ambitious and departs substantially from teachers' existing practice, as is the case in efforts to promote standards-based mathematics instruction such as the effort described in this chapter (EEPA 1990; Cohen and Hill 2001; Spillane, Reiser, and Reimer 2002; Thompson and Zeuli 1999).

Conceptualizing the organizational changes that must take place to support teachers' learning has received even less attention. Intervening at the organizational level is important to consider alongside teacher-level interventions, because research has shown that conditions in schools often work against opportunities for teachers to learn new approaches and sustain research-based interventions. Teachers usually work in isolation with little opportunity to share their practice with others (Little 1990; Lortie 1975). Although teachers' professional communities can be sites for teachers' learning (Franke and Kazemi 2001; Little 1982, 2003; McLaughlin and Talbert 2001; Smylie and Hart 1999), they are often not present in most schools. When they are present, they do not always support teachers' improvement (Gallucci 2003; Stein and Coburn 2008). Finally, principals typically do not engage in and support teachers' efforts to learn new, ambitious forms of instruction; even principals who want to engage teachers in instructional conversations usually are not equipped to do so deeply or meaningfully (Coldren and Spillane 2007).

In this chapter, we argue that the promise of tools as vehicles for supporting large-scale, research-based improvement will remain unfulfilled without parallel attention to teachers' learning and the organizational supports required for it. Most tool designers concentrate their efforts on students' learning; we have focused on the design and study of tools that can support research-based teachers' learning and organizational change. Because prior research suggests the promise of lesson planning as an essential feature of ambitious instruction (Perry and Lewis 2010), we decided to use a research-based lesson-planning tool as the primary catalyst for schoolwide instructional improvement. Although we designed this tool for day-to-day use by *individual teachers*, it also serves an important role at *professional learning community* and *leadership* levels. By serving as an anchor for new, more instructionally focused routines at these levels as well, the lesson-planning tool can be viewed as a boundary object (Wenger 1998), intended to coordinate and align practices across teacher, teachers' professional community, and school levels.

This chapter unfolds in three sections. First, we describe the overall project that includes our work. Then we turn to a description of the ways in which we integrate research and practice into our work, illustrating the research-practice links across all three of our work's levels. We conclude with a discussion of the implications of our work for tool designers who aim to use research to improve schools and classrooms.

The Project

Our effort is part of a larger project to design and study an approach to schoolwide reform that is based in a theory of change that integrates attention to instructional improvement and school-level organizational change. The project is housed within a partnership between university-based researchers and public schoolteachers at a grades 6–10 public school for traditionally underserved youth. The university-based team includes people with organizational and school reform expertise as well as instructional experts in mathematics, science, social studies, and English language arts. Although our work spans four subjects, this chapter focuses on the mathematics work. The school-based team consists of teachers, teacher leaders, an instructional coach, and two principals.

Methodologically, the project is committed to a design-based process that develops, tests, implements, and revises tools and routines as they grow in scope and complexity. The design task that we have carved out for ourselves consists of iterative development of routines, rooted in lesson planning, that guide professional work at the school, teachers'-professional-community, and individual-teacher levels. We focus on routines because they are widely acknowledged to be powerful mechanisms that shape how organizations accomplish work and, as such, introducing new routines may be a source of organizational change (Feldman 2000; Feldman and Pentland 2003). We are anchoring the routines in lesson planning because research on teaching identifies crucial distinctions between how expert and novice teachers plan lessons (Leinhardt 1993) and suggests that lesson planning may be a high-leverage strategy for effecting change in teachers' practice (Lewis and Tsuchida 1998; Perry and Lewis 2010; Watanabe 2002; Yoshida 1999).

Our primary objective is to create a viable prototype of a systemic, integrated set of routines rooted in lesson planning that improves (*a*) the school organization by orienting principals' work and routines anchored in instruction and learning; (*b*) the quality of professional conversations in teachers' professional communities by introducing new routines of interaction around lesson planning, observation, and reflection; and (*c*) the rigor, coherence, and evidence base of instructional practice through deprivatizing, deepening, and systematizing the ways in which teachers prepare for and deliver lessons.

How Our Work Integrates Research and Practice

As illustrated in figure 3.1, research and practice interact in *two* distinct ways in our work. First, extant research knowledge is embedded in the initial iteration of the tools and routines at each level. Hence, tools serve as vehicles to carry research into practice.

Second, our own project-based research on how educators actually enact the routines in practice (*a*) leads to improvements in the tools and routines (see feedback circle in fig. 3.1) and (*b*) feeds into the more generalized research base on teachers' improvement in mathematics education, teachers' professional community, and organizational routines that support instructional improvement (see the bottom right-hand corner of fig. 3.1). Each of these functions of our project-based research bridges research and practice, albeit in different ways. The "improvement" focus bridges them by using research to create an optimally functioning tool for practice in the

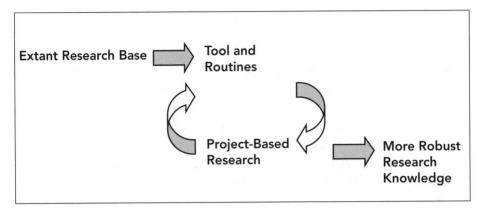

Fig. 3.1. Interactions between research and practice in the project

local environment. The "generalized research" focus does so by documenting and analyzing improvement efforts and then framing the findings in a theoretically rich set of concepts and ideas. This latter focus helps build fundamental understanding of school improvement processes by elaborating the extant research base. It also helps transfer research knowledge to new sites by providing conceptual structure that can support future improvement efforts. (Because we were in the earliest stages of the project when we wrote this chapter, we do not discuss the development of generalized knowledge about school improvement here.) As described below, these pathways between research and practice occur at all three levels of the work: individual teachers, teachers' professional communities, and the school as an organization.

Research-Practice Links for Individual Teachers

Extant research. In order to introduce the lesson-planning focus at the very beginning of the 2009–2010 school year, the design team created the initial version of the lesson-planning tool (LPT). This tool drew heavily from the Teaching through a Lesson protocol (Smith, Bill, and Hughes 2008), which also came from research. In response to earlier research demonstrating the difficulties that teachers have enacting ambitious mathematics lessons (i.e., lessons composed of tasks that are student-centered, open-ended, and cognitively demanding) (Stein, Grover, and Henningsen 1996; Stigler and Hiebert 2004), Smith and her colleagues identified elements of instructional practice that correlate to productive use of students' thinking during such lessons, fundamental features of which were planning effectively in general and anticipating students' thinking in particular (Stein et al. 2008). Accordingly, we designed the LPT as an online template that prompts teachers to (*a*) set clear goals for students' learning, expressed as important mathematical ideas, (*b*) select or adapt appropriately challenging instructional tasks aligned with those goals, (*c*) anticipate different ways that students might approach the task, and (*d*) plan appropriate ways to support students' thinking regardless of their approach to the task (see fig. 3.2).

With the agreement of the district and the teachers' union, project teachers' weekly postings of LPT-based lesson plans to an electronic repository, which school principals could also access, replaced the district's existing routine that all teachers turn in paper lesson plans each Friday. Typically, these paper plans took a daily "block" form and consisted of brief notes identifying what part of the curriculum the teachers would cover. The notes, often written as page numbers

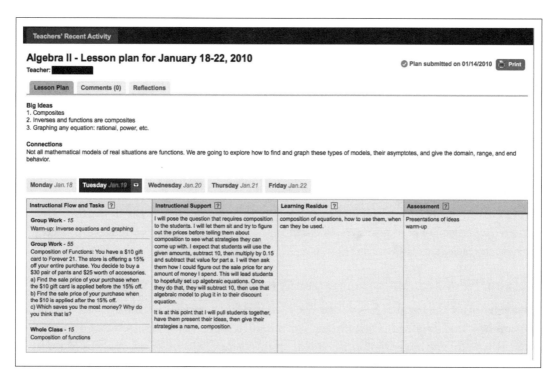

Fig. 3.2. Lesson planning template (designed by Jennifer Cartier, Jennifer Russell, Margaret Smith, and Mary Kay Stein)

or section headings of textbooks, focused on what students would *do* during various activities, not on how students would *think* about the ideas or problems with which they would be presented. Although this new form of planning demanded more time, we believed that using the LPT would lead to improved instruction, because completing it requires teachers to imagine and plan for the contingencies that arise during lessons, especially lessons that focus on higher-level, cognitively demanding tasks. Moreover, we hoped that the repository would eventually become a shared database of lesson plans that all teachers would value as useful starting points for future planning. The shared database would resemble what is available—albeit in print form—in other countries and a nascent example of what Hiebert, Gallimore and Stigler (2002) have called a *professional knowledge base for teaching*.

Project-based research. As we write this chapter, we are halfway through the first year of a three-year project. As such, our research on how teachers use the LPT is in its earliest phases. We have, however, encountered an early challenge: teachers have difficulty seeing the connection between their planning and their ability to enact better lessons. Our baseline interviews with teachers offer some insight into why teachers might not naturally consider lesson planning a useful activity for improving instructional practice. Most teachers said that, in the past, they undertook lesson planning only because someone else mandated it, not because it would be personally useful.

The designers, however, see planning and the improvement of practice as integrally interrelated. Lessons, the designers argue, are more effective when (*a*) they are designed with a clear mathematical goal in mind, and (*b*) teachers have prepared by anticipating the many routes, both correct and incorrect, that students may take toward that goal. Without this planning, the designers argue, teachers either (*a*) resort to not taking students' thinking into account, simply taking over the thinking and telling students what to do; or (*b*) become hostage to an "anything goes" approach to dealing with students' thinking.

Teachers' skepticism regarding the usefulness of planning with the LPT led us to change the tool's design. We added a "reflection prompt," which encouraged teachers to think about their enacted lessons in light of their plans. We urged teachers to go back to their lesson plans daily after teaching each lesson, in order to assess whether they had adequately anticipated the many ways that students approached the task and the extent to which students learned the important mathematical ideas. We hoped that taking this extra step—of reflecting on a lesson in light of one's lesson plans—would more firmly cement the connection between the two. After introducing reflection prompts to teachers at a faculty meeting, we planned to carefully track how much teachers used them and how.

Research-Practice Links in Teachers' Professional Communities

Extant research. The research base for the LPT guided how teachers *should* plan, but it did little to show how teachers might *learn* to plan. For the latter, we turned to research that demonstrates that—for any innovation to take hold—teachers need ongoing, work-embedded assistance (Garet et al. 2001). To design this assistance, we drew on research suggesting that teachers' professional communities could potentially support teachers' ambitious learning (Smylie and Hart 1999). We also drew on research that suggests that—without firm, expert guidance—interactions in teachers' professional communities can, and often do, go no further than story swapping, sharing materials, and providing discrete bits of information or advice (Lortie 1975; Little 1990; McLaughlin and Talbert 2001). Thus, our initial design called for dedicated, weekly planning sessions attended by university partners and teachers in each subject. We would anchor these sessions—at least initially—in planning. Placing our bets with a lesson-planning anchor for the early work of the teachers' professional communities also drew on past research—in this instance, research that has demonstrated that more in-depth interactions can replace shallow routines through an intensive focus on lesson planning (Perry and Lewis 2010).

Project-based research. Preliminary analyses of these weekly meetings' field notes suggested that teachers were often reluctant to focus on lesson planning during this "dedicated" time because of the urgency of other concerns (e.g., scoring formative assessments, preparing materials aligned with the state assessment). Also, teachers continued to have difficulty seeing how this new, enhanced form of lesson planning paid off in better lessons. Even when we modified the routine from the expectation that each individual teacher would plan on his or her own to that of everyone taking part in joint planning around a single lesson, the teachers continued to struggle.

At this point, the university-based partners planned an intervention that focused squarely on the intersection of planning and practice. One of the partners, Anna, completed the LPT for a lesson that used the Orange Juice Task (Lappan et al. 2009). Next, Anna taught the lesson to students in the classroom of one of the project teachers, and she videotaped the lesson. Afterward, the teachers and university-based partners together reviewed Anna's lesson plan and watched a clip from the videotape. The teachers were amazed to see a fundamental student misconception that Anna had anticipated in her lesson plan actually played out during the video-taped lesson. An especially rich discussion began when Anna admitted that, although she could anticipate both the difficulties students might have and the resulting questions she might ask when this occurred, she could not move the students toward correct understandings during the lesson. This discussion raised questions that the teachers had not considered about students' difficulties with proportional reasoning. At the end of the session, one of the teachers asked if they could spend future sessions discussing students' possible solutions for, and misconceptions of, a task that they would be teaching in the upcoming week.

Overall, this experience seemed to produce a renewed desire to focus on planning tasks as a group. A new routine (i.e., plan for lessons by anticipating how students will approach instructional tasks; have one teacher teach and videotape the lesson; reflect on the videotape as a group, discussing what their joint planning had and had not prepared them for) therefore appears to be in the making, one that could potentially increase the value that teachers ascribe to lesson planning. Moreover, this approach allows teachers to undertake the work as a group first. Joint planning not only offsets the cost of time required to complete the LPT, but also leads to discussions anchored by a concrete artifact (the lesson plan) and to shared investment in the joint product (the lesson).

Research-Practice Links in the School

Extant research. The research on teachers' learning provides guidance for how teachers learn to enact innovations, but it offers little advice regarding how one might intervene at the organizational level to support further, or at least not detract from, that learning. For the latter advice, we turn to research that has identified routines as central to the way organizations work (Cyert and March 1963; March and Simon 1958) and to the ways organizations learn (Levitt and March 1988). Typically, schools as organizations do not welcome innovations aimed at the core of teaching and learning, at least partially because the organizational routines of schooling tend to center on noninstructional tasks, such as tracking students' attendance, allocating resources, accomplishing yearly budgeting cycles, and maintaining hiring practices (Conley and Enomoto 2005).

An emerging body of research, however, has documented how routines may drive planned change in schools (Coburn and Russell 2008; Resnick and Spillane 2007; Sherer and Spillane, forthcoming). This research suggested to us that our design needed to find a way to incorporate instructional improvement into the school's organizational routines; one way to do this, we reasoned, would be to develop the school principals' instructional leadership capacities. This can be a daunting task, however: research identifies several obstacles that confront principals who attempt to become instructional leaders. The biggest obstacle is figuring out how to connect to instructionally meaningful issues and events (Coldren and Spillane 2007). Other than by observing classrooms, many school leaders are at a loss for how to enter into an "instruction and learning" conversation.

Following Coldren and Spillane (2007), who suggest that tools can serve as boundary objects for helping principals make the connection to instruction and learning, we worked with one school leader to develop a routine that involved reviewing teachers' lesson plans as they appeared in the LPT. This review was part of a larger routine of cycles of lesson observations, feedback to teachers, and tailored professional development. We viewed the insertion of lesson plans into the larger cyclical routine as an opportunity (*a*) to stress the connections between high-quality lesson plans and effective lessons and, by so doing, expand educators' views of the practice of teaching to include planning, enactment, and reflection; and (*b*) to give leaders a window into teachers' practice based on a concrete artifact (i.e., the lesson plan) that is already part of both the teachers' and the principal's work. Now, however, the artifact's use stresses learning, whereas before it stressed compliance.

As with the lesson-planning routine, we attempted to co-opt an existing routine—in this instance, the Learning Walk—to new ends with the introduction of new, more demanding tools. The Learning Walk, the signature tool of the Institute for Learning at the University of Pittsburgh (www.instituteforlearning.org), is a district-level routine that occurs three or four times a year. The Institute describes a Learning Walk as an organized visit through a school's hallways and classrooms, during which participating district leaders and curriculum specialists, or "walkers," collect evidence of teaching, learning, and how teaching affects learning. Walkers visit a school for half a day, during which they converse with school leaders and observe classrooms. At the end of the Learning Walk, walkers look for patterns in the school and think about next steps needed to support teachers as they continue to refine instruction. The Learning Walk's stated purpose is to get a sense of instructional quality in the school and, more important, to learn about the school's plans for improving instructional quality.

Recognizing the Learning Walk's potential as practiced by the district, the principals and university partners decided to adopt their own, ongoing cycles of observations, feedback to teachers, and professional development as a way to improve instruction continually. They designed each cycle to include three to five teachers. A cycle would begin with the principal and the university partners reviewing the lesson plans of the teachers that they were going to observe. Next, they would conduct observations lasting at least thirty minutes each. Afterward, the university partners and principal would debrief in the principal's office, coming to agreement on the strengths and weaknesses of each teacher's instruction and next steps for that teacher's professional development. Within twenty-four hours, the principal and the university partners present the teachers with feedback and make firm plans for followup professional development. The process would hold accountable the university partners, for providing the professional development; the principal for making sure that teachers attended and took the professional development seriously; and the teachers, for trying their best to integrate learnings from the professional development into their classroom practice.

Project-based research. When we were writing this chapter, we had conducted and studied one cycle of this new organizational routine. Already, some positive outcomes had occurred. For example, the feedback that one teacher received showed her that, despite intentions to the contrary, her instructional practice had devolved primarily into low-level, procedural tasks. Shortly thereafter, she agreed to coplan a lesson that would challenge her students more.

Improvements in the routine, however, remain to be made. In particular, the design team needs to devise ways to help the principal see the connection between lesson planning and instructional quality. Initially, the principal did not think to include lesson-plan review in the larger leadership routine. Once reviewing teachers' lesson plans became a part of the principal's routine, she used them to monitor instructional fidelity, that is, she would check to see if the teachers were on the topic that they said they would be on. Given principals' normal role of enforcing compliance with lesson-plan submission and typically spending little time examining the plans' content and quality, the principal interpreting her role as compliance enforcement was very predictable. We should have anticipated it.

Our plan moving forward is to develop tools that more explicitly draw links between the level of students' thinking and reasoning observed during classroom lessons and the level of planning in which the teacher has engaged. Our first draft consists of a rubric for distinguishing richer versus more impoverished plans. We are now working on an observational protocol that can be used alongside that rubric. We look forward to future discussions with the principal, using these tools to look across teachers' plans and their instructional practices to find integrated solutions for moving forward.

Discussion

This chapter focuses on a primary reason for lack of research uptake in educational practice: limited awareness of—and thus limited use of—the multiple pathways and mechanisms that interconnect research and practice. Focusing on tools as the bridging mechanism, we have given examples of two pathways between research and practice. The first is embedding extant research knowledge in tools that teachers use in practice; the second concerns the role and contributions of project-based research in efforts to improve instructional practice.

We are not the first to suggest tools as a way to bridge research and practice. However, our chapter complicates the notion of how tools bridge research and practice for effective, school-wide instructional improvement. Designing tools that embed research results is the easy part. Assuring that schools use those tools in ways that will improve schoolwide instruction and increase students' learning is more challenging. We have framed that challenge, at least in part, as a need to attend to issues of teachers' learning.

Even when the ultimate goal is to promote students' learning, teachers are the vehicles for reaching students. Our chapter suggests that an intentional focus on teachers' learning is crucial if schools expect teachers to integrate new, research-based instructional approaches in ways that support teachers' learning goals for students. As the project teachers' skepticism of the LPT's usefulness demonstrated, without attention to teachers' learning needs and supports that make sense in their local circumstances, teachers can easily not use tools or use them only partially or superficially.

Yet, designers are rarely explicit about the demands new tools place on teachers' learning. As noted earlier, when designers do focus on teachers, they aim for how teachers *should* teach, rather than how teachers might *learn how to teach* in ways that research suggests would benefit students (Fishman and Davis 2006). By illustrating the challenge of teachers not seeing the connection between lesson planning and improved instruction, this chapter suggests that explicit attention to how teachers learn is an important part of ensuring that research-based approaches actually make it into classroom practice.

We should note that the project discussed in this chapter did anticipate the need to attend to teachers' learning. For example, the project dedicated time for coplanning with university-based partners. Nevertheless, our initial research only took us so far. Design-based research—our continual monitoring of how teachers were and were not using the LPT to plan lessons—is what alerted us to the need to refine not only the tool, but also the routines surrounding the tool. This insight suggests the importance of continual research on local enactments of tool-based innovations, in order to assure that the innovations' users make specific and timely adjustments. The bottom line is that extant research only gets one so far.

Most design efforts marginalize issues of organizational or systemic change to support teachers' learning even further. Yet, this chapter suggests that the practice side of the research-practice equation matters for research and development, and it matters significantly. Most designers of tool-based educational innovations have struggled with implementation because of contextual conditions in schools or districts. The question is how to intervene at the organizational level to create more favorable conditions. We began by focusing on one aspect of the school's organizational environment—instructional leadership. Simultaneously, other parts of the partnership focused on districtwide constraints. Even so, other constraints in the system were equally or more pressing (e.g., a high-stakes state test that focuses primarily on procedural knowledge and skills). Unless the project continually monitors these constraints and intervenes appropriately and at appropriate times, the constraints may swamp our fledgling school improvement. The bottom line here is that, if an organizational constraint places an undue burden on project teachers or pulls them in an opposite direction, then our improvement efforts will not flourish.

Our parting message is that tool designers should, as much as possible, create designs that address the context of schools and classrooms as well as teachers' learning needs. If designers pay attention to more than the instructional tool, and if they also consider ways to foster environments in schools that encourage teachers to learn new approaches that the tools carry, we will have come a long way.

REFERENCES

Atkinson, Richard C., and Gregg B. Jackson, eds. *Research and Education Reform: Roles for the Office of Educational Research and Improvement.* Washington, D.C.: National Academies Press, 1997.

Ball, Deborah Loewenberg, and David A. Cohen. "Reform by the Book: What Is—or Might Be—the Role of Curriculum Materials in Teacher Learning and Instructional Reform." *Educational Researcher* 25 (December 1996): 6–8, 14.

Brown, Anne, James Greeno, Magdelene Lampert, Hugh Mehan, and Lauren B. Resnick. *Recommendations Regarding Research Priorities: An Advisory Report to the National Educational Research Policy and Priorities Board.* Washington, D.C.: National Academy of Education, 1999.

Burkhardt, Hugh, and Alan H. Schoenfeld. "Improving Educational Research: Toward a More Useful, More Influential, and Better Funded Enterprise." *Educational Researcher* 32 (December 2003): 3–14.

Coburn, Cynthia E., and Jennifer L. Russell. "District Policy and Teachers' Social Networks." *Education Evaluation and Policy Analysis* 30 (September 2008): 203–35.

Coburn, Cynthia E., and Mary Kay Stein, eds. *Research and Practice in Education: Building Alliances, Bridging the Divide.* Lanham, Md.: Rowman and Littlefield, 2010.

Cohen, David K., and Carol A. Barnes. "Pedagogy and Policy." In *Teaching for Understanding: Challenges for Policy and Practice,* edited by Milbrey McLaughlin and Jan Talbert, pp. 207–39. San Francisco: Jossey Bass, 1993.

Cohen, David K, and Heather C. Hill. *Learning Policy: When State Education Reform Works.* New Haven, Conn.: Yale University Press, 2001.

Coldren, Amy F., and James P. Spillane. "Making Connections to Teaching Practice: The Role of Boundary Practices in Instructional Leadership." *Educational Policy* 21 (May 2007): 369–96.

Conley, Sharon, and Ernestine K. Enomoto. "Routines in School Organizations: Creating Stability and Change." *Journal of Educational Administration* 43 (January 2005): 9–21.

Cyert, Richard M., and James G. March. *A Behavioral Theory of the Firm.* Englewood Cliffs, N.J.: Prentice-Hall, 1963.

EEPA. *Educational Evaluation and Policy Analysis (EEPA)* 12 (Fall 1990): 233–353.

Fishman, Barry J., and Elizabeth A. Davis. "Teacher Learning Research and the Learning Sciences." In *The Cambridge Handbook of the Learning Sciences,* edited by Ronald K. Sawyer, pp. 535–50. Cambridge: Cambridge University Press, 2006.

Feldman, Martha S. "Organizational Routines as a Source of Continuous Change." *Organization Science* 11 (November 2000): 611–29.

Feldman, Martha S., and Brian T. Pentland. "Reconceptualizing Organizational Routines as a Source of Flexibility and Change." *Administrative Science Quarterly* 48 (March 2003): 94–118.

Franke, Megan, and Elham Kazemi. "Teaching as Learning within a Community of Practice: Characterizing Generative Growth." In *Beyond Classical Pedagogy in Teaching Elementary Mathematics: The Nature of Facilitative Teaching,* edited by Terry Wood, Barbara Nelson, and Janet Warfield, pp. 47–74. Mahwah, N.J.: Lawrence Erlbaum Associates, 2001.

Gallucci, Chrysan. "Communities of Practice and the Mediation of Teachers' Responses to Standards-Based Reform." *Education Policy Analysis Archives* 11, no. 35 (2003). Retrieved 9/29/03 from http://epaa.asu.edu/epaa/v11n35.

Garet, Michael, Andrew Porter, Laura Desimone, Beatrice Birman, and Kwang Suk Yoon. "What Makes Professional Development Effective? Results from a National Sample of Teachers." *American Educational Research Journal* 38 (Winter 2001): 915–45.

Hiebert, James, Ronald Gallimore, and James W. Stigler. "A Knowledge Base for the Teaching Profession: What Would It Look Like, and How Can We Get One?" *Educational Researcher* 31 (June–July 2002): 3–15.

Kaestle, Carl F. "The Awful Reputation of Educational Research." *Educational Researcher* 22 (January–February 1993): 23–31.

Kennedy, Mary M. "The Connection between Research and Practice." *Educational Researcher* 26 (October 1997): 4–12.

Lappan, Glenda, James T. Fey, Elizabeth D. Phillips, William M. Fitzgerald, and Susan N. Friel. *Comparing and Scaling: Ratio, Proportion and Percent.* Upper Saddle River, N.J.: Prentice-Hall, 2009.

Leinhardt, Gaea. "On Teaching." In *Advances in Instructional Psychology,* Vol. 4, edited by Robert Glaser, pp. 1–54. Hillsdale, N.J.: Lawrence Erlbaum Associates, 1993.

Levitt, Barbara, and James G. March. "Organizational Learning." *Annual Review of Sociology* 14 (August 1988): 319–38.

Lewis, Catherine, and Ineko Tsuchida. "A Lesson Is Like a Swiftly Flowing River: Research Lessons and the Improvement of Japanese Education." *American Educator* 22 (Winter 1998): 14–17, 50–52.

Little, Judith Warren. "Norms of Collegiality and Experimentation: Workplace Conditions of School Success." *American Educational Research Journal* 19 (Fall 1982): 325–40.

———. "The Persistence of Privacy: Autonomy and Initiative in Teachers' Professional Relations." *Teachers College Record* 91 (Summer 1990): 509–36.

———. "Inside Teacher Community: Representations of Classroom Practice." *Teachers College Record* 105 (August 2003): 913–45.

Lortie, Daniel C. *Schoolteacher: A Sociological Study.* Chicago: University of Chicago Press, 1975.

March, James, and Herbert Simon. *Organizations*. New York: J. Wiley and Sons, 1958.

McLaughlin, Milbrey W., and Joan E. Talbert. *Professional Communities and the Work of High School Teaching*. Chicago: University of Chicago Press, 2001.

Perry, Rebecca, and Catherine Lewis. "Building Demand for Research through Lesson Study." In *Research and Practice in Education: Building Alliances, Bridging the Divide,* edited by Cynthia E. Coburn and Mary Kay Stein. Lanham, Md.: Rowman and Littlefield, 2010.

Resnick, Lauren B., and James P. Spillane. "From Individual Learning to Organizational Designs for Learning." In *Instructional Psychology: Past, Present and Future Trends,* Advances in Learning and Instruction series, edited by Lieven Verschaffel, Fillip Dochy, Monique Boekaerts, and Stella Vosniadou, pp. 259–76. Oxford, U.K.: Elsevier, 2006.

Sherer, Jennifer Zoltners, and James P. Spillane. "Constancy and Change in Work Practice in Schools: The Role of Organizational Routines." *Teachers College Record,* forthcoming.

Smith, Margaret S., Victoria Bill, and Elizabeth Hughes. "Thinking through a Lesson Protocol: A Key for Successfully Implementing High-Level Tasks." *Mathematics Teaching in the Middle School* 14 (October 2008): 132–38.

Smylie, Mark A., and Anne W. Hart. "School Leadership for Teacher Learning and Change: A Human and Social Capital Development Perspective." In *Handbook of Research on Educational Administration,* edited by Joseph Murphy and Karen Seashore Louis, pp. 421–41. San Francisco: Jossey-Bass Publishers, 1999.

Spillane, James P., Brian J. Reiser, and Todd Reimer. "Policy Implementation and Cognition: Reframing and Refocusing Implementation Research." *Review of Educational Research* 72 (Fall 2002): 387–431.

Stein, Mary Kay, and Cynthia E. Coburn. "Architectures for Learning: A Comparative Analysis of Two Urban Districts." *American Journal of Education* 114 (August 2008): 583–626.

Stein, Mary Kay, Barbara Grover, and Marjorie Henningsen. "Building Student Capacity for Mathematical Thinking and Reasoning: An Analysis of Mathematical Tasks Used in Reform Classrooms." *American Educational Research Journal* 33 (Summer 1996): 455–88.

Stein, Mary Kay, Randi A. Engle, Margaret S. Smith, and Elizabeth K. Hughes. "Orchestrating Productive Mathematical Discussions: Helping Teachers Learn to Better Incorporate Student Thinking." *Mathematical Thinking and Learning* 10 (October 2008): 313–40.

Stigler, James W., and James Hiebert. "Improving Mathematics Teaching." *Educational Leadership* 61 (February 2004): 12–16.

Thompson, Charles L., and John S. Zeuli. "The Frame and the Tapestry: Standards-Based Reform and Professional Development." In *Teaching as the Learning Profession: Handbook of Policy and Practice,* edited by Linda Darling-Hammond and Gary Sykes, pp. 341–75. San Francisco,: Jossey-Bass, 1999.

Watanabe, Tad. "Learning from Japanese Lesson Study." *Educational Researcher* 59 (March 2002): 36–39.

Wenger, Etienne. *Communities of Practice: Learning, Meaning and Identity*. Cambridge: Cambridge University Press, 1998.

Yoshida, Makoto. "Lesson Study: A Case Study of a Japanese Approach to Improving Instruction through School-Based Teacher Development." Ph.D. diss., University of Chicago, 1999.

Building Bridges between Research and the Worlds of Policy and Practice: Lessons Learned from PROM/SE and TIMSS

William H. Schmidt

IN 2003, the National Science Foundation (NSF) awarded Michigan State University (MSU) a $35 million dollar grant for the Promoting Rigorous Outcomes in Mathematics and Science Education (PROM/SE; MSU n.d.) project under the auspices of the Mathematics and Science Partnership program. The money would support a partnership including sixty-one local districts from Ohio and Michigan. The organizational structure, however, was not directly through the districts themselves, but involved five partners that were aggregates of a number of districts.

The two partners in Ohio were nongovernmental consortia in the Cleveland and Cincinnati regions. These consortia represented the local school districts in those two metropolitan areas that volunteered to participate. The two organizations, known as SMART and High AIMS, founded prior to the NSF grant's awarding, represented constituencies whose aim was improving mathematics and science learning, supported in part through businesses in their communities. The SMART consortium came into being from participation in the Third International Mathematics and Science Study (TIMSS). The 1995 TIMSS included an international curriculum analysis that extensively coded textbooks and national standards of some fifty countries including the United States. In the 1995 TIMSS study, the SMART consortium was one of the few state or local organizations that could participate as an entity comparable to a "country."

The three partners in Michigan were formal, state-level organizations known as *intermediate school districts*, each of which supported the local school districts in its county. The three intermediate districts that participated in PROM/SE were Calhoun, Ingham, and Saint Clair counties.

In many ways, the sixty-one districts under these five partnerships represented a microcosm of the United States. The districts included large, urban areas such as Cleveland and Cincinnati; midsized cities such as Lansing, Michigan; numerous suburbs surrounding the metropolitan areas; and rural areas, some impoverished. These districts' demographics taken together paralleled those of the United States as a whole. During the project's initial phase, we administered the

1995 TIMSS items and those of other tests to students in the sixty-one districts, in order to get a baseline measure of achievement. The baseline measures allowed us to compare these students' achievement results to those in the 1995 TIMSS U.S. sample. We found no major departures from the national norms on any of the basic indicators.

We established PROM/SE with the goal of using the TIMSS results to improve students' learning in mathematics and science in these sixty-one districts. Toward that end, PROM/SE first concentrated on helping the districts formulate coherent, rigorous, focused curricular standards and approaches to mathematics and science in the contexts of Michigan's and Ohio's state standards. We defined a curriculum as coherent when it articulated its topics over time in a sequence consistent with the logical and, if appropriate, hierarchical nature of the discipline's content from which the subject matter derives (Schmidt, Houang, and Cogan 2002; Schmidt, Wang, and McKnight 2005). The second focus was building a means to increase teachers' content knowledge. Several other, related goals existed, such as eliminating tracking. The organizational structure included the two Ohio consortia, the three Michigan intermediate school districts, and MSU. This core group collaboratively designed and developed the strategies that would address these goals.

What follows will explain how that partnership could link together research, policy, and practice. Since the 1995 TIMSS formed this work's intellectual foundation and instrumentation, this chapter will also draw on that TIMSS experience in addressing this volume's central question.

The Research Base

The PROM/SE research's intellectual underpinnings derived directly from TIMSS, not only in research findings that influenced the PROM/SE's conceptual framework and goals, but also in the project's actual research work. PROM/SE's conceptualization had research playing a central role, not just in the typical NSF sense of evaluating the project's effectiveness, but as an integral part of the work itself.

In the study's base year, using modified TIMSS instrumentation, we developed tests and had some 200,000 grades 3–12 students throughout the sixty-one districts take them. We developed three tests, one each for grades 3–5, 6–8, and 9–12. The tests incorporated the relevant 1995 TIMSS items, but an additional set of items were also developed. The result totaled 465 mathematics items for elementary school, 345 for middle school, and 450 for high school. We created similar tests for science, but from here on, this chapter will only address the mathematics part of the study. We based the testing strategy on the duplex design (Bock and Mislevy 1988), which allowed us to specify 22, 28, and 26 content strands at each of the three broad grade-level spans. We used the three tests to assess students' knowledge of very specific mathematics topics. We thereby identified the weak areas in students' performance, one indication of the content on which we needed to focus.

Also, we developed detailed questionnaires, parts of which relied heavily on the original 1995 TIMSS instrumentation, for teachers, principals, and district leaders. Most important was the teachers' questionnaire, which included a section detailing the amount of instructional time teachers spent teaching each of some thirty to forty mathematics topics, depending on the grade levels. Teachers estimated the number of periods over the school year that they spent teaching each particular topic. Another part of the teachers' questionnaire reported the degree to which the teachers believed that they had the appropriate academic background for teaching each specific topic. In all instances, we adapted the instrumentation specifically to the grade levels.

We asked principals questions about their schools and the schools' focus on instructional issues. District leaders answered questions about the districts' content standards, textbooks used, and their policies with respect to curriculum, including tracking. Together, the data from the principals' and district leaders' questionnaires gave us an understanding of district curricular policies and helped us specify what mathematics topics each grade level intended to cover. Teachers, however, reported what curricular topics they actually implemented in each of their classrooms. Combined, the data from all three questionnaire types provided information on content coverage or students' opportunity to learn (OTL) for each of the 8000 classrooms in the study. Combined with students' achievement data, this information provided the districts and the project a rich knowledge base from which to develop plans of action. These plans would focus on those topics where the data analysis correlated students' poor performance with teachers' lack of adequate background. The data collection's longitudinal aspects allowed us to tailor those plans each year on the basis of the previous year's data.

From the data analysis, we identified several problematic topics for each of the three broad grade levels. We then designed capacity-building (professional development) activities around coherence to address content knowledge in those topics, which included the following.

- Elementary school (grades K–5): Fractions. geometry and measurement, whole numbers, number and operations, number theory, angles, area and perimeter, and algebraic reasoning

- Middle school (grades 6–8): Geometry and measurement, equations and lines, rational numbers, fractions and decimal operations, angles, proportionality, number theory, area and perimeter, and algebraic reasoning

- High school (grades 9–12): Mathematics of change, data and chance, algebra, geometric thinking, number theory, angles, and area and perimeter

The districts received the data about OTL, students' performance, and teachers' self-reported knowledge enthusiastically, because they had never before had such specific information. District leaders and teachers found the reports based on those data extremely helpful as they looked at their own standing as an organization. The strength of these data for our work, and for that of the districts individually, was how specific they made the achievement results and curriculum descriptions. For example, seven test areas for fractions alone in grades 3–5 enabled the districts to identify very specific strengths and weaknesses in fractions.

This database was an essential, central feature of the project, always the guiding focus for developing capacity-building activities and designing particular interventions. We repeated the achievement tests in the project's fourth, fifth, and sixth years. When we were writing this chapter, we had one additional annual data collection planned. We collected teachers' and districts' data each year of the project except for year two. This collection gave us longitudinal data at the individual, classroom, school, and district levels, allowing us to track changes over time.

Elsewhere in the project, we designed and implemented an experimental study that randomly allocated districts to a fractional factorial design, a one-eighth replicate of a 2^6 factorial design. That study, designed in the project's third year, enabled us to look at the impact that working with districts on both micro (within a grade) and macro (between grades) coherence had on students' learning.

PROM/SE's work centered on research throughout the partnership. This somewhat unique aspect, we believed, contributed to integrating research, practice, and policy. The research base for PROM/SE's work did not derive from other projects or other studies, except initially, when TIMSS research influenced the study's conception. Instead, the research base derived from the research embedded in and part of the partnership's *zeitgeist* (i.e., its intellectual and cultural climate). This derivation gave the districts a sense of ownership as they received district-specific reports from all data collections. The research base gave leadership data and a sense of direction from which to make instructional changes. The research base also gave the districts data on the strengths and weaknesses not only in students' performance, but also in the curriculum's coherence, focus, and rigor, both as the district intended it and as the teachers implemented it.

Bumps along the Road

What this chapter has reported so far seems to imply a very smooth connection between research and the districts' practice. That implication warrants some additional comments. Collecting extensive data, as just described, was a burden for the districts. Teachers often met the collection with cries of "why do we have to do this," even though the five partners supported the effort and worked very hard to secure the participating districts' cooperation. Testing all grades 3–12 students in all the districts was a tremendous burden; getting all the teachers in the districts to fill out the questionnaires was no less arduous. In all, we collected data from more than 200,000 students, and more than 7,000 teachers participated in some aspect of the project.

The collection task was inordinately large and difficult to pull off. Resistance was very strong at times, especially in the first year, when we had to collect the baseline data. Given the short time we had to obtain the data (i.e., some five months after the start of the project), we found that MSU initially created many problems by not anticipating individual districts' needs and difficulties. So, when we came back to the districts for the next data collection some two years later, strong resistance reflected frustration, voiced and unvoiced, at the extra time and commitment needed, at many of the districts' levels, for the intensive data collection. Such problems and an unwillingness to participate in these large efforts occurred at every phase of data collection, although it decreased as the research team became more efficient and sensitive to the districts' needs.

All this has implications for integrating research and practice, especially because it relates to the collaboration between those of us responsible for schooling and those responsible for designing and carrying out the research. The two groups had different goals, and they tended to impede each other at times. The data-collection burden was the major impediment to cooperation. Other contributing difficulties included what appeared to us, as researchers, to be simple issues, such as scheduling activities or teachers and administrators setting low priority to cooperating with us. These issues, too, affected how rigorous and scientific the program design and data collection could be.

What ultimately helped overcome these difficulties and accomplish PROM/SE's work was the close working relationship that developed between the five partners—the three intermediate districts in Michigan and the two consortia in Ohio—and MSU. The five partners helped translate the goals and research requirements from the project team to the districts. Since the partners already had good working relationships and consistent interests with their districts, they could obtain greater cooperation and facilitate the project's goals. Those of us who designed this work and carried out the data collections believe strongly that without the five partners' participation as go-betweens, the project would have struggled. Without them, we could not have secured the cooperation needed from the districts to collect such extensive research data. Having the five partners' cooperation therefore was invaluable and crucial to the project's success. Perhaps the more general principle that might emerge from this experience is that having an intermediate entity, closer to the world of practice but not directly engaged in that practice, can greatly facilitate collaboration in conducting the research. Even more generally, such an intermediary can facilitate integrating research and practice.

Policy Levers

The discussion so far has not dealt directly with the third aspect of the triad—the world of policy. Research findings influenced work with teachers on developing coherence in their classrooms—the world of practice. In the world of policy, we worked at three levels—school, district, and state. At all three levels, we focused on using the knowledge gained, both from the research and from working with the teachers on capacity building related to the development of coherence, to recognize the appropriate policy levers and to develop recommendations regarding what the policies should be.

At the school level, we focused on the principals, helping them understand that teachers' time was the most valuable resource for which a principal is responsible. How teachers allocate that time to teaching various topics in a subject is crucial to learning. The central element of our focus was helping principals understand that two of their major responsibilities were (1) to work within the district as instructional leaders, and then (2) to help develop strategies for accomplishing teachers' effective use of time to support coherent instruction. Here the major lever was school policy regarding the principal observing classrooms and monitoring teachers' time allocations for curriculum topics.

At the district level, we worked directly with superintendents and curriculum directors to design strategies to achieve coherence in the mathematics curriculum. Involving both the district leaders and the principals in this process was absolutely essential. Initially we did not take that possible collaboration into account. By examining relevant data, we discovered that the resulting discussions among district leaders and principals were crucial to success. Therefore, we increased the amount of attention given to coordination between the two groups. We learned early in the project that teachers cannot accomplish the projects' goals without support from the principals and the district superintendents. Integrated working across these levels, among small groups of teachers, principals, and district curriculum leaders, made developing a coherent curriculum easier.

Our final policy lever involved state leaders. Early on, we included the state leadership in both Michigan and Ohio, among them the superintendents of education for both states. This move directly affected both states' policies. For example, a committee that included PROM/SE members developed Michigan's grades K–8 standards. Ongoing PROM/SE work also influenced the standards, along with the international benchmarking developed as part of the original TIMSS study. This input directly resulted in draft Michigan standards, which were only slightly modified as they progressed through the typical channels necessary for adopting state standards. A PROM/SE principal investigator also played a central role in developing Michigan's high school standards. The project maintained close contact with Ohio's state officials, too, and influenced the current revisions of their standards in both science and mathematics.

Here this chapter will take the liberty to incorporate some lessons learned as a part of involvement in TIMSS. After we had assembled the results and written the reports, we had the opportunity to carry the work's policy implications to state and national levels. We spoke with many of the governors, chief state school officers, and state legislators who debate and make educational policy. The international benchmarking work related to the curriculum put us in the position of being able to bring to the table data that addressed directly the issue that concerned governors: how their standards compared to those of other countries, especially the high-achieving countries. The presence of an extensive database, and an understanding of that database's implications for policy, served as the entrée into the world of policymakers, who were greatly interested in obtaining such insights to structure their policies meaningfully. Just as was true in bridging research to practice, the availability of high-quality, broadly representative data relevant to important issues greatly influenced the bridge to policy.

What We Have Learned about Bridging Research, Practice, and Policy

This section summarizes what we have learned from both TIMSS and PROM/SE. Through those two projects, we have come to a better understanding of how one can connect research to both practice and policy. The summarized lessons are as follows:

- Practice and policy communities must recognize the research as relevant, large-scale, and of high quality both conceptually and statistically. Policy leaders must believe that what you have is solid evidence, relevant and important for them to pay attention to. Achieving this description is especially difficult in the world of practice, where one wants to influence what teachers and administrators actually do in the classrooms and in the schools. Here was where we had the greatest difficulty. What helped was figuring out ways to present the data and findings straightforwardly, with a minimum of the technical detail usually found in academic research. Teachers and principals had to understand clearly what these data implied about their particular situation and the practices in which they engaged. We found that teachers and principals could not typically draw implications from the data tables and displays. The research team had to point out the explicit connections to the teachers' practices. At PROM/SE's outset, we operated under the premise that "if we built it, they would know how to use it." Such was not the case. We therefore needed to provide training on how to interpret the data.

- The intermediate level of organization between researchers and schools, as described previously, helped especially in obtaining schools' and districts' greater cooperation. In PROM/SE, the two consortia in Ohio and the three intermediate school districts in Michigan filled that role. Such organizations have a long history of working with districts and have earned districts' trust. Working through such intermediaries may not be possible in all situations. Where such entities are available, however, we would strongly recommend engaging them as important organizational features that facilitate bridging research and practice.

- One of the difficulties working with teachers concerning content knowledge in mathematics is that "they often don't know what they do not know, and as a result, they do not know what they need to know." Teachers' levels of mathematical knowledge in PROM/SE were such that they didn't recognize the value of the capacity building around mathematical coherence because in their minds this would not help them in their classrooms Monday morning. Overcoming this view was difficult and took time and effort.

- Districts usually are not organized particularly well for engaging in major research. Although the district leaders endorsed PROM/SE's goals and committed to them at least in principle, the actual work's realities simply became just one more distraction from the leaders' heavy workloads. No one ever explicitly expressed such to us, but it was clear in many ways that district leaders gave what we were asking them to do a low priority. This implied a deeper principle: most likely a clash is inevitable between the research culture, where you constantly modify the plan on the basis of what you learn from data, and the schooling culture, where plans are set quite far in advance and changes often cause conflicts or cannot be accommodated.

- The factor that made the biggest difference in our success in working with schools was the local-level leadership. The districts in which we accomplished the most were those with strong district leaders and principals. These people saw their role as that of leadership, especially in instructional and curricular issues.

- A serious impediment to bridging research and practice is that the primary players—the district and school leaders—change faster than longer-term projects take to carry out their work and achieve desired results. By the fourth year of PROM/SE, almost half the districts had superintendents different from those in place when the districts committed to the project initially. These differences caused further problems, because those new superintendents were not necessarily as committed to the project's work as their predecessors were.

- A communication strategy guided and designed by standard principles of public relations facilitates access to the policy community. Although such a strategy is not necessary, its availability tends to make access easier by making the results more visible. The TIMSS project engaged a public relations firm that helped make the results more visible, enabling greater access to different levels of the policy community.

Bridging the worlds of practice, research, and policy does not happen by chance and will not happen by itself. Instead, it requires considerable effort. Projects that desire an influence beyond producing a research report need to recognize this difficulty; build into the project's timeline the amount of time that such a bridging takes; have adequate, needed resources available; and have the patience to work through all this. Without that bridging, research projects will produce reports read by academics rather than something that actually influences practice and policy.

REFERENCES

Bock, R. Darrell, and Robert J. Mislevy. "Comprehensive Educational Assessment for the States: The Duplex Design." *Educational Evaluation and Policy Analysis* 10 (Summer 1988): 89–105.

Michigan State University (MSU). Promoting Rigorous Outcomes in Mathematics and Science Education (PROM/SE) Project, East Lansing, Mich.: PROM/SE, n.d. Retrieved December 4, 2009, from http://www.promse.msu.edu/.

Schmidt, William H., Hsing Chi Wang, and Curtis C. McKnight. "Curriculum Coherence: An Examination of U.S. Mathematics and Science Content Standards from an International Perspective." *Journal of Curriculum Studies* 27 (September 2005): 525–59.

Schmidt, William H., Richard Houang, and Leland Cogan. "A Coherent Curriculum: The Case of Mathematics." *American Educator* 26 (Summer 2002): 13–28.

Teachers' Use of Standards-Based Instructional Materials: Partnering to Research Urban Mathematics Education Reform

Karen D. King

THROUGH its Discovery Research K–12 program in 2008, the National Science Foundation (NSF) funded a multiyear project, Teachers' Use of Standards-based Instructional Materials (TUSIM) that, at the time of this writing, researchers are completing at New York University (NYU), Newark (N.J.) Public Schools (NPS), and the Quality Education for Minorities (QEM) network. We are studying how teachers use the mandated instructional materials, Connected Mathematics Project (CMP), in urban schools. The mixed-methods study includes case studies of individual teacher's use of CMP in two schools, population surveys of all middle grades mathematics teachers in the NPS district, and policy analyses that describe the creation and nature of the district's instructional regime. This approach's intents are to help (1) us as researchers understand the impact of mandated instructional policies and (2) the local school district's administrators assess such policies' effect on instruction. The aim is for the study's results to guide future policymaking and practice.

This collaboration builds on a successful Local System Change (LSC) project that NPS undertook from 2002 to 2008, funded by NSF grant ESI-0138806. During that project, NPS adopted a set of NSF-funded textbooks for the elementary and middle grades and conducted ongoing professional development for teachers to support those materials' implementation. During that implementation, NPS collaborated with Rutgers University and others to set up specific professional development, such as that conducted in the district's after-school faculty sessions. This chapter's study expands NPS's collaborators in mathematics education to include

This material is based on work supported by the National Science Foundation under grant no. DRL-0732184. Any opinions, findings, conclusions, or recommendations expressed in this material are those of the author and do not necessarily reflect the views of the National Science Foundation. The author would like to thank her collaborators, Ognjen Simic, Candace Barriteau Phaire, Mellie Torres, Jessica Tybursky, and the teachers and administrators of Newark, N.J., Public Schools, for their contributions.

NYU and QEM Network. We also use results from the previous evaluation studies of the district (Gearheardt et al. 2009; Samuel 2009; TCC Group 2007), which drew influence largely from NSF's evaluation of the overall LSC program.

The Partners

Three colleagues, with different experiences and expertise in mathematics education, initiated and supported this collaboration: May Samuels, of NPS; Monica Mitchell, initially with QEM Network and later of MERAssociates (MERA); and Karen King, of New York University. We three came together through the LSC program at NSF. Bringing different strengths and experiences from academia, school districts, and the nonprofit education sector, we collaborated to develop and execute a research agenda that would both add to our existing knowledge about curriculum implementation in an urban district and guide future directions for NPS's implementation. Having created such a mutually beneficial research study supports our continued work together.

The Newark Public Schools enrolls 42,149 students, which places the district in the group of 1.9 percent of U.S. school districts with enrollments more than 25,000. Such districts educate a total of 16 million students, 34.1 percent of the students enrolled in U.S. public schools (NCES 2007). The largest school district in New Jersey, NPS serves a diverse population: approximately 59 percent are of African descent (African American, Caribbean, and West African); 31 percent, labeled Hispanic heritage (Puerto Rican, Latin American, and South American); 8 percent, classified as white; and 1 percent, Asian or other heritage. NPS has identified 41 languages other than English spoken in its students' homes. During the 2007–08 school year, the district classified 9 percent of its students as English language learners (ELLs), and 14 percent received special education services. Seventy-seven percent of the students are from families whose income qualifies them for free or reduced lunch programs, and 40 percent enter or leave their school after the school year has begun. Thus, the NPS fairly represents large, urban districts around the country, although with fewer students classified as ELLs than the 14-percent national average (Keigher 2009).

NYU is the largest private nonprofit university in the United States. It is a research-intensive university with a global presence. NYU's Steinhardt School of Culture, Education, and Human Development is a professional school that houses eleven departments with more than 180 full-time faculty and 6500 undergraduate and graduate students. Mathematics education is part of the Department of Teaching and Learning, which includes a core focus on research in urban settings and teachers' education. Over the TUSIM project's course, the project's staff has worked with the Center for Research on Teaching and Learning, directed by Robert Tobias, and included other full-time faculty, doctoral students, and master's students.

The QEM Network is a nonprofit organization in Washington, D.C., dedicated to improving education for minorities throughout the nation. Employing a networking and coalition-building approach, QEM Network is as a national resource and catalyst to help unite and strengthen educational restructuring to benefit minority children, youth, and adults. The network also advances minority participation and leadership in the national debate on how best to ensure access to a quality education for all citizens. MERA is an independent education research-and-evaluation consultancy specializing in science, technology, engineering, and mathematics education. It also deals with issues of access and equity that broaden participation for underrepresented students and communities across the grades K–16 continuum. Working with school districts, universities, not-for-profits, and foundations, MERA brings expertise in bridging research and practice based on proven experience in large-scale systemic reform, multipartner collaborations across disparate sectors, and leveraging interventions as intermediaries between academia and grades K–12 schooling.

The TUSIM Project

The TUSIM project studies the ways that middle grades teachers use NSF-funded, Standards-based mathematics instructional materials (NCTM 1989). NPS mandated the materials' use and provided teachers with a wide range of curriculum-focused professional development opportunities. In most urban school systems, district administrators with responsibilities for mathematics education face immense challenges. They must adopt and implement, on a large scale, policies that increase students' mathematical knowledge—the larger the student and teacher populations, the more challenging the implementation. These district administrators have few policy options under their control. Also, mathematics supervisors and other district administrators with responsibilities for mathematics education have little to no control over the quality of the teaching force who will implement locally mandated or developed policies. Often, for example, state offices set standards for certification and human resources offices or building administrators handle hiring decisions. A district's mathematics administrators thus have little sway over the qualifications and quality of mathematics teachers that the district hires. Yet, these mathematics administrators must ensure that teachers provide high-quality mathematics instruction to all students.

The TUSIM project created an opportunity for the university-based research team to conduct a study consistent with a research-intensive university's mission. At the same time, Samuels and NPS stood to benefit from understanding, through a fine-grained analysis, (1) how their teachers were using the instructional materials and (2) what supports and barriers existed to standards-based instruction and students' learning. Mitchell, a former principal investigator for an LSC in New York City and a former program officer for NSF's LSC program, served as both project manager and boundary spanner. She attended to the sometimes-competing needs and goals of the research team and school district leaders. A boundary spanner, who can focus on the project without pressures from institutional constraints at NYU and NPS, crucially supports our research while maintaining that research's connection to both policy and practice.

Research Base

Our collaboration drew on several lines of research: curriculum as a policy lever for scaling up reform-based instructional practice (Confrey and Stohl 2004), particularly allowing for the content's alignment to the state standards and assessment (Smith and O'Day 1990; Porter and

McMaken 2010); curriculum policy as part of a larger instructional regime (Ball and Cohen 1996); and our hypothesized theory of action for how we intend curriculum to affect students' learning, ultimately through sequencing instructional tasks with appropriate levels of cognitive demand (Stein, Grover, and Henningsen 1996; Henningsen and Stein 1997; Lappan, Phillips, and Fey 2007). As Wilson (2003) describes in her book on the mathematics education reforms of the 1990s in California, the state instructional materials adoption process is crucial, because the adoptions create a vision of what mathematics instruction policymakers intend to implement in classrooms. Although we focused our work at the district level, the district's systemic change effort also gave us an opportunity to explore how teachers' practice aligned with the district's curricular vision as described in the district's policy documents, particularly its curriculum pacing guide.

Although most instructional materials developers have a curricular vision (Brown et al. 2009), many school districts convey their own vision through a curriculum guide, scope, sequence, or framework linked to the instructional materials. To address alignment and confront instructional realities of the NPS' instructional regime (Mitchell and King 2010), the district's pacing guide modified and adapted the sequencing and use of units proposed by the authors of the Connected Mathematics Project. Examining surveys focused on (1) the specifics of how teachers used the curriculum and (2) the Surveys of Enacted Curriculum (SEC; see Porter and McMaken [2010]), we found significant diversion by many teachers from the district's vision as stated in the curriculum pacing guide. For example, 71 percent of the 159 middle grades mathematics teachers surveyed reported using materials other than CMP in their teaching. Of those teachers, 81 percent reported using these alternative materials—modifications that could potentially change the district's curricular vision—at least once a week.

When surveyed about unit-by-unit and lesson-by-lesson using CMP's implementation survey, TUSIM's teachers reported a range of use of the materials, from fully complying with the district's curriculum pacing guide to directly contradicting it. Tables 5.1 and 5.2 present data on the percent of teachers, by grade level of the class they were teaching, who reported using a unit as written, using it with some adaptation, or not using it. Table 5.1 presents those units that both the pacing guide and CMP's implementation information suggest using in a given grade, along with a suggested order. Table 5.2 presents those units the curriculum pacing guide either moves from the recommended grade or rearranges order in ways that affect data collection. For example, the district pacing guide moves two units, Variables and Patterns and Moving Straight Ahead, from seventh grade, as the curriculum developers designed, to eighth grade in an effort to focus on algebra in eighth grade. However, as noted in table 5.2, 93 percent of seventh-grade mathematics teachers reported using Variables and Patterns as is or modified, and 34 percent reported using Moving Straight Ahead as is or modified. Although this diversion from the materials' suggested use is not evidence that the teachers do not align their work with the district's specified mathematical content and broader vision, it does call into question how valid the idea of focusing on textbook fidelity would be as a method of scaling up the district's curricular vision.

At the same time, we found that research did not clearly articulate exactly how instructional materials affect students' learning. We adapt from Stein, Remillard, and Smith (2007) a vision of curricular impact on students' learning that makes clear that the teachers' instruction is a

Table 5.1

Modification Levels for Units Aligned with Pacing Guide and Curriculum Implementation Suggestions

Gr.	Unit Title	CMP version	N	Used As Is (%)	Adapted or One of Many (%)	Replaced or Did Not Use (%)
6	Prime Time	1 & 2	62	39.5	58.3	2.2
6	Bits and Pieces I	1 & 2	61	31.8	66.8	1.4
6	Shapes and Designs	1 & 2	61	36.6	56.8	6.6
6	Data about Us	1	15	44.0	48.0	8.0
6	How Likely Is It	2	45	28.3	41.7	30.0
6	Bits and Pieces II	2	44	27.8	62.5	9.7
7	Stretching and Shrinking	1 & 2	46	31.3	60.5	8.2
7	Comparing and Scaling	1 & 2	45	28.1	64.8	7.1
7	Accentuate the Negative	1	14	10.0	70.0	20.0
7	Filling and Wrapping	1 & 2	46	12.2	28.9	58.9
7	Data Distributions	2	32	23.4	33.6	43.0
8	Thinking with Mathematical Models*	1 & 2	52	30.0	37.2	12.3
8	Looking for Pythagoras	1 & 2	52	19.2	57.0	23.8
8	Kaleidoscopes, Hubcaps and Mirrors	2	35	38.3	49.1	12.6

*Does not add to 100%

Table 5.2

Modification Levels for Units Moved and Not Taught According to Pacing Guide

Gr.	Unit Title	CMP version	N	Used As Is (%)	Adapted or One of Many (%)	Replaced or Did Not Use (%)
6	Bits and Pieces II	1	17	31.4	60.8	7.8
6	Bits and Pieces III	2	45	10.8	24.8	64.5
7	Variables and Patterns	1 & 2	46	29.6	63.8	6.6
7	Moving Straight Ahead	1 & 2	46	12.0	22.4	65.6
7	Bits and Pieces II	1	14	8.3	89.3	2.4
7	Bits and Pieces III	2	32	31.7	53.9	14.4
8	Growing, Growing, Growing	1	14	1.8	51.8	46.4
8	Frogs, Fleas, and Painted Cubes	1 & 2	46	6.0	17.9	76.1
8	Say it with Symbols	1 & 2	46	10.0	23.0	67.0
8	Samples and Populations	1	15	5.0	58.3	36.7
8	Clever Counting	1	14	5.7	51.4	42.9

Table 5.2.—*Continued*

Gr.	Unit Title	CMP version	N	Used As Is (%)	Adapted or One of Many (%)	Replaced or Did Not Use (%)
8	The Shapes of Algebra	2	32	10.0	15.6	74.4
8	Variables and Patterns	1 & 2	52	24.8	56.8	18.4
8	Accentuate the Negative	1	15	18.7	61.3	20.0
8	Moving Straight Ahead	1 & 2	52	25.3	67.6	7.1

significant mediator between instructional materials and students' learning (see fig. 5.1). We define instructional materials as those that provide a series of tasks sequenced for students' learning and based on understandings of particular learning theories. We seek to understand what adaptations teachers make to the tasks in planning, including replacing the tasks entirely. Also, we hope to understand how teachers use the tasks to alter cognitive demand in an effort to influence students' learning positively. Both the qualitative case studies and quantitative survey data collection were designed to increase our understanding of this important interaction. This inquiry advanced two interdependent endeavors. First, the research contributes to scholarship related to curricular impact on students' learning. Second, the partnership creates a means for feeding our research findings to school district leadership. NPS can therefore focus professional development on primary aspects of teachers' planning and instruction to support the district's curricular vision. The partnership's work links research to policy with an intended impact on practice.

Results of the Collaboration

The ongoing collaboration has produced results at different levels. First, the data collected coupled with our analysis give Samuels a basis for making revisions to the NPS's curriculum pacing guide (see Mitchell and King [2010]). The pacing guide was new for the second edition of CMP. However, data collected about developing teachers' practices and aligning them with the pacing guide pointed to the need to evaluate and modify the pacing guide in the next year.

Second, while seeking an explanation for the previously presented results on teachers' textbook use across the units described in tables 5.1 and 5.2, we found curriculum articulation to be an area of concern. Specifically, we identified pressures on the eighth-grade mathematics curriculum from high school initiatives related to algebra readiness. We hypothesize that the reason some teachers do not adhere to the curriculum pacing guide during seventh- and eighth-grade algebra units is probably because they feel pressure to prepare students for high school algebra.

Third, at the teachers' level, the immediate feedback they receive from the SEC, and their ability to continue to use such data for planning, allow the teachers to reflect on an entire year of instruction. Teachers noted that although participating in the surveys was demanding, the professional development we provided prior to data collection was very useful. It focused on attending to the cognitive demand of the tasks used in teaching, prompting teachers to reflect on their teaching in ways not previously considered. Many reported that their individual results

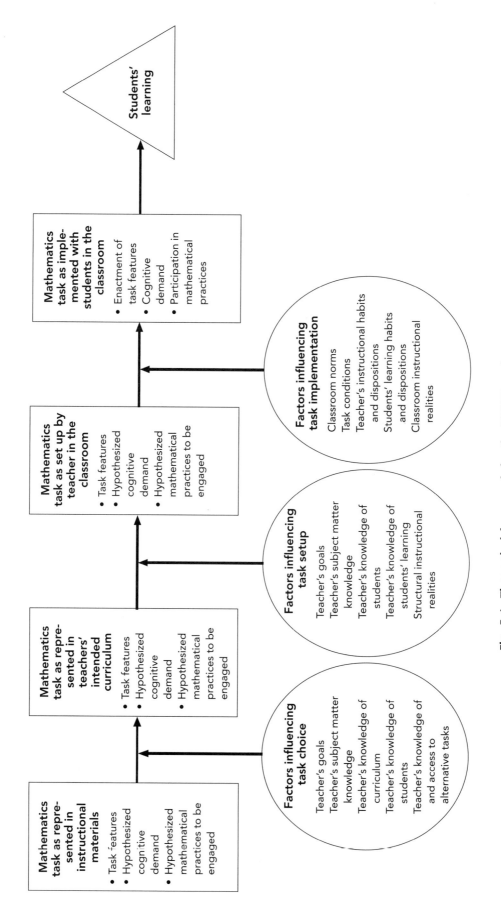

Fig. 5.1. Theoretical framework for the TUSIM project

Mathematics task as represented in instructional materials
• Task features
• Hypothesized cognitive demand
• Hypothesized mathematical practices to be engaged

Factors influencing task choice
Teacher's goals
Teacher's subject matter knowledge
Teacher's knowledge of curriculum
Teacher's knowledge of students
Teacher's knowledge of and access to alternative tasks

Mathematics task as represented in teachers' intended curriculum
• Task features
• Hypothesized cognitive demand
• Hypothesized mathematical practices to be engaged

Factors influencing task setup
Teacher's goals
Teacher's subject matter knowledge
Teacher's knowledge of students
Teacher's knowledge of students' learning
Structural instructional realities

Mathematics task as set up by teacher in the classroom
• Task features
• Hypothesized cognitive demand
• Hypothesized mathematical practices to be engaged

Factors influencing task implementation
Classroom norms
Task conditions
Teacher's instructional habits and dispositions
Students' learning habits and dispositions
Classroom instructional realities

Mathematics task as implemented with students in the classroom
• Enactment of task features
• Cognitive demand
• Participation in mathematical practices

Students' learning

were surprising considering the narrowness of some of the tasks they posed in their classrooms. The teachers suggested that the results motivated them to be more attentive to the demands of the tasks they present, particularly the tasks' breadth and range.

Results of the data collection, analysis, and interpretation reflect the close relationship between research and practice envisioned in the design of data collection. For example, before the teachers participated in the survey, the researchers conducted a professional development session on the cognitive demands of tasks. This exercise enabled researchers to gain the trust of teachers and demonstrate a desire to understand the teachers' perspectives and instructional realities in using CMP. We believe that the trust engendered during the professional development motivated teachers to articulate concerns and challenges faced when implementing mandated curricula in special education. Our emerging results indicate that special-education teachers are significantly more likely to modify the text than their general-education colleagues.

In our original work plan, we proposed offering several professional development workshops for the teachers, on the basis of their results on both the CMP survey and the SEC. However, recent changes in district leadership led to a freeze on professional development provided by external groups. We remain hopeful that we can resume the professional development in the near future. Although the teachers participating in this study have had many opportunities for professional development focused on curriculum implementation, we envision offering sessions that allow teachers to reflect on their own practice at a different scale, across an entire year of instruction.

Our continued analysis intends to link results of students' achievement to teachers' self-reported practice on the two surveys through multilevel modeling. We anticipate that as we learn more from our results, we will continue to give the district feedback, primarily through Samuels and her policymaker role as the district's mathematics supervisor, to help shape future professional development initiatives.

Observations and Cautions

On the basis of our experiences, we offer several observations for those considering ways to connect research, policy, and practice. A crucial element in advancing our work's collaboration is the contribution of a boundary spanner who stands outside the university and school district contexts but has intimate knowledge of both. Mitchell's role has been essential in facilitating data collection and analysis, as well as in negotiating the varying needs of both Samuels, at the district, and King, at the university. At the same time, other necessary components are partnerships with a researcher, who can conceptualize the research study, plan for the types of data needed, and oversee analysis; and with a high-level district administrator, who has sufficient experience to negotiate the district culture and offer crucial insights on conducting the research. Samuels's role as an administrator, having both the authority to make decisions and an understanding of classroom practice, was a very important element in the research and in communicating findings to teachers. Given the administrative instabilities in many urban districts, Samuels's long tenure in NPS as the district mathematics coordinator, and her credibility with teachers, served our project well.

We have one cautionary note, however, related to leadership instability in many urban districts. District leadership changes curtailed Samuels's authority to manage mathematics professional development, thereby halting our ongoing professional development around the SEC data available to teachers. Although this situation did not affect the ongoing collaboration, our ability to influence practice through ongoing professional development for teachers was severely limited. We look forward to resuming that relationship in the near future.

Aligning the research questions and methods so that they guide practice and policy is a important result from our collaboration. Although broader research into curriculum implementation, and the theoretical and empirical questions therein, framed our research, we also organized data collection and analysis to answer practical questions that would be useful to the district. The latter presents an ethical concern in that, in the context of the district, findings from our data-gathering activities might be interpreted negatively. Thus, we take care to report on teachers in ways that respect their confidentiality. Further, Samuels, in particular, has to tread a fine line between her dual roles as researcher and district leader and the sometimes-competing ethical demands of the two. To address this dilemma, we organize our analyses so that they report Samuels's participation on blinded data, such as transcripts, and with nonidentified and nonidentifiable teachers' data. Collaborations such as ours should plan for such ethical problems from the start and seek to shield district personnel and research participants from exposure to them. The Institutional Review Board (IRB) process should work to ensure that the research plans accounts for any ethical issues.

Finally, our partnership has thrived in a district with a culture of working with and relying on research and evaluation in decision making for mathematics curriculum and instruction. Because of Samuels's commitment to research and evaluation in her own work, the teachers' past experiences participating in research and evaluation studies, and, most important, the district mathematics office's willingness to implement new policies derived from the research and evaluation in which its administrators and teachers participated, conducting our research was easier than it might have been. The district's IRB process was clear and smooth. At the same time, we had to be flexible in working with the district and respect the demands of what we were asking of teachers. We had to plan data collection around state test administration. After initially collecting data from most teachers, we had to have a makeup day for those teachers who otherwise could not attend the first professional development session. We carefully ensured that the data collection and our instruments would spark reflection among teachers, to serve their ends as well as our research goals. Our willingness to be flexible around the district's needs gave us a richer data set than we might have had in other circumstances.

In conclusion, although difficulties exist in conducting research that guides and is guided by policy and practice, the rewards are great. The opportunity to link research and practice benefits practitioners by involving them in research that focuses on concerns encountered in their daily practices. Collaborating with practitioners who have highlighted important constructs that we had not previously considered deeply enriched our research. One example has been the role of the district's curricular vision, as embodied in the curriculum pacing guide, as a pivotal mediator for teachers' use of mathematics instructional materials. Such collaboration also highlights essential practical issues needing further study, such as the role of the special-education mathematics teacher in mathematics curriculum reform. Collaborations and partnerships among school districts, schools, and universities, mediated by invested boundary spanners, support a strategic focus on linking research, policy, and practice in order to enhance mathematics education.

REFERENCES

Ball, Deborah Loewenberg, and David K. Cohen. "Reform by the Book: What Is—or Might Be—the Role of Curriculum Materials in Teacher Learning and Instructional Reform?" *Educational Researcher* 25 (December 1996): 6, 8, 14.

Brown, Stacy A., Kathleen Pitvorec, Catherine Ditto, and Catherine Randall Kelso. "Reconceiving Fidelity of Implementation: An Investigation of Elementary Whole-Number Lessons." *Journal for Research in Mathematics Education* 40 (July 2009): 363–95.

Confrey, Jere, and Vicki Stohl, eds. *On Evaluating Curricular Effectiveness: Judging the Quality of K–12 Mathematics Evaluations.* Washington, D.C.: National Academies Press, 2004.

Gearhart, Darleen L., Roberta Y. Schorr, Lisa B. Warner, and May L. Samuels. *Standards-Based Curricula, Professional Development, and Student Achievement on New Jersey Assessments: Evaluation of the Effects on Student Achievement in Grades K through 8 of the Local Systemic Change Project, NPSSIM.* Newark, N.J.: Newark Public Schools, 2009.

Henningsen, Marjorie, and Mary Kay Stein. "Mathematical Tasks and Student Cognition: Classroom-Based Factors That Support and Inhibit High-Level Mathematical Thinking and Reasoning." *Journal for Research in Mathematics Education* 28 (November 1997): 524–49.

Keigher, Ashley. *Characteristics of Public, Private, and Bureau of Indian Education Elementary and Secondary Schools in the United States: Results from the 2007–08 Schools and Staffing Survey.* Washington, D.C.: U.S. Department of Education, National Center for Education Statistics, 2009. Retrieved June 21, 2010, from http://nces.ed.gov/pubs2009/2009321.pdf.

Lappan, Glenda, Elizabeth Difanis Phillips, and James T. Fey. "The Case of 'Connected Mathematics.'" In *Perspectives on the Design and Development of School Mathematics Curricula,* edited by Christian Hirsch, pp. 67–79. Reston, Va.: National Council of Teachers of Mathematics, 2007.

Mitchell, Monica, and Karen D. King. "The Development of a Mathematics Curriculum Pacing Guide in an Urban School District." Paper presented as part of a symposium at the Research Presession of the National Council of Teachers of Mathematics. San Diego, Calif., April 2010.

National Center for Education Statistics (NCES) Common Core of Data. "Local Education Agency Universe Survey," 1979–80 through 2005–06. Washington D.C.: United States Department of Education, 2007.

National Council of Teachers of Mathematics (NCTM). *Curriculum and Evaluation Standards for School Mathematics.* Reston, Va.: NCTM 1989.

Porter, Andrew, and Jennifer McMaken, and Rolf K. Blank. "Surveys of Enacted Curriculum and the Council of Chief State School Officers Collaborative." In *Disrupting Tradition: Pathways for Research and Practice in Mathematics Education,* edited by William F. Tate, Karen D. King, and Celia Rousseau Anderson, pp. 21–31. Reston, Va.: National Council of Teachers of Mathematics, 2010.

Samuels, May L. "Final Report for Elementary, Secondary, and Informal Education: Newark Public Schools Systemic Initiative in Mathematics." Newark, N.J.: Newark Public Schools, 2009. ESI-0138806.

Smith, Marshall S., and Jennifer O'Day. "Systemic School Reform." *Journal of Education Policy* 5, no. 5 (1990): 233–67.

Stein, Mary Kay, Barbara W. Grover, and Majorie Henningsen. "Building Student Capacity for Mathematical Thinking and Reasoning: An Analysis of Mathematical Tasks Used in Reform Classrooms." *American Educational Research Journal* 33 (July 1996): 455–88.

Stein, Mary Kay, Janine T. Remillard, and Margaret S. Smith. "How Curriculum Influences Student Learning." In *Second Handbook of Research on Mathematics Teaching and Learning,* 2nd ed., vol. 1, edited by Frank K. Lester, Jr., pp. 319–69. Charlotte, N.C.: Information Age Publishing and National Council of Teachers of Mathematics, 2007.

TCC Group. "Newark Public Schools' Local Systemic Change Initiative Core Evaluation Report: Year 5, 2006–2007." New York: TCC Group, 2007.

Wilson, Suzanne M. *California Dreaming: Reforming Mathematics Education.* New Haven, Conn.: Yale University Press, 2003.

Examining What We Know for Sure: Tracking in Middle Grades Mathematics

Lee V. Stiff
Janet L. Johnson
Patrick Akos

TRACKING has been used throughout educational systems for decades, ostensibly to benefit both the students who are capable of more rigorous coursework and students who might find such work too challenging. Many educators believe that to do otherwise would be a disservice to both groups of students. Proponents of tracking insist that by separating the classrooms into slow and fast learners, teachers can better focus their instruction and adjust the pace to match students' abilities (Hopkins 2009). This chapter will chronicle one large school district's challenges and successes in connecting research, practice, and policy to meet the mathematics education needs of its students. Lessons learned—or failed—should help other school districts better understand what they know and don't know about access to high-quality mathematics.

Lately, more and more research indicates that although tracking benefits those placed in the higher tracks, many students placed in the lower tracks find themselves in an educational downward trajectory that can detrimentally affect their education, and ultimately, their lives. After students have been tracked, they usually remain in the high or low track where they were initially placed, and the achievement gaps between the two become greater over time (Wheelock 1992; O'Connor, Lewis, and Mueller 2007). In fact, evidence suggests that removing tracking and teaching all students as if they were high achievers does not "drag down" high achievers, but rather pulls up the performance of average students (Garrity 2004). Indeed, assigning students to lower-ability groupings depresses students' learning regardless of students' ability levels (Hallinan 2003).

Few would argue that students who are tracked high in math will learn more rigorous math content than they would if they were tracked low. But, what about the students who are tracked low? What quality of math content do they learn? And, if we do track students, what should the criteria for high and low tracks be? Indeed, should students ever be tracked out of high-quality math offerings?

One group of researchers analyzed data to determine the effects of tracking on students with similar math abilities and found that when high average (C+) students were placed in low-, middle-, and high-track courses in middle school, the percents of students who successfully completed two college-prep math classes in high school were 2 percent for the low track, 23 percent for the middle track, and 91 percent for the high track (Burris, Heubert, and Levin 2006). In other words, students' placement in the top math track created greater success for students in their future schooling.

But, more surprising, even when content mastery is a criterion for determining course placements, other criteria are frequently used to make tracking decisions. In fact, research has shown that non-Asian minorities are less likely to be placed in higher-level courses than other, equally qualified students (Vanfossen, Jones, and Spade 1987; O'Connor, Lewis, and Mueller 2007). For example, Stone and Turba (1999) reported that in one California school district, of the students who demonstrated the ability to be admitted into algebra, only 51 percent of the blacks and 42 percent of the Latinos were admitted, whereas 100 percent of the Asians and 88 percent of the whites were.

In one statewide study, educators who did not have easy access to academic achievement data to determine how to track students unabashedly admitted to using demographic factors to make such decisions. Specifically, school counselors, when asked if they used academic data when advising students for course enrollment or academic interventions, reported that they used academic and behavioral data to inform themselves about how to serve students better. However, when asked to describe that data, many described free or reduced lunch status as academic data. A few school counselors reported that, because they lack demographic data about socioeconomic status, they had to rely on race to identify students who had barriers to learning and therefore would benefit from counseling services or other referrals for interventions (Johnson et al. 2005).

Clearly, it appears that students' ethnicity or economic status can trump performance data when one decides how to place students into high-quality courses. In all likelihood, using a combination of demographic factors makes it possible for educators unknowingly to track students by race. Undoubtedly, mathematics placement decisions based on such demographic information negatively affect, if not create, the achievement gap and diminish future learning opportunities.

Otherwise-capable students placed into low math tracks have shown a decrease in their mathematics self-efficacy (Akos, Shoffner, and Ellis 2007; Callahan 2005). Combined with inappropriate instruction and teachers' beliefs about social barriers and the lack of support, students face the prospects of lowered expectations and the resulting lower grades that ultimately affect their long-term college and career choices. Unfortunately, this scenario has disproportionately affected minorities (Akos, Shoffner, and Ellis 2007). The deleterious effects of lower self-efficacy can cause anxiety, which has been shown to affect cognition and performance neurologically (Gray, Braver, and Raichle 2002). In other words, when capable students are placed into low math tracks, their performance gets worse. The fact of the matter is, all students benefit from taking rigorous coursework (Hallinan 2003). We know this; yet too often, we ignore it. In fact, although an association exists for all students between taking rigorous high school mathematics and going to college, the relationship is even greater for students whose parents' education did not go beyond high school (Choy 2002).

Tracking students has at least two major flaws. First, tracking has little or no benefit for students, especially those placed in the lower tracks. Second, tracking disproportionately affects minority and low-socioeconomic-status students irrespective of their prior achievements. A standard review, conducted by the school's counseling department of a large school district in North Carolina, noticed the flaws associated with tracking in mathematics. The counseling department realized that school counselors could not properly perform their duties without the benefit of relevant information about students and cooperation among school counselors, math teachers, and administrators. The department recognized that it needed a venue to bring school personnel together. So, in response to the need, the department created a forum of experts in school counseling, mathematics education, and the proper use of data, which would address tracking issues in middle and high school.

The forum, known as the School Counseling/Math Collaborative, examined existing students' performance and access realities of course taking in mathematics; best practices in teaching math and school counseling, and the need to share them; and the need to learn, develop, and use research-based school counseling practices. Moreover, the collaborative addressed the need for objective criteria for identifying students who would likely succeed in rigorous math courses. In addition to and in support of the collaborative, EDSTAR Analytics, an education consulting and evaluation firm, conducted Data Academies for the school district. Each Data Academy worked with school-based teams consisting of school counselors, math teachers, and administrators to help them use data to improve students' performance and close achievement gaps in mathematics. Each Data Academy typically had a cohort of seven schools.

A Data Academy confronted issues about using data, both subjective and objective, to understand how best to serve the needs of all students. Specifically, each Data Academy addressed issues related to instructional practices, school and district policies affecting math achievement and access, and school personnel's attitudes and beliefs about students' performance in mathematics.

The Data Academies supported the school district's study for two years. In the first year, Data Academy facilitators met twice each month in full-day sessions with planning teams from the district to review how best to support the schools in the district. Also, each school-based team met one day each month with Data Academy facilitators and received on-site support twice a month. Early in the second year, school-based teams met with their Data Academy facilitators for one full day to review their schools' data and implementation plans. Subsequently, Data Academy facilitators furnished onsite support once a quarter or as needed.

What follows recounts many of these Data Academies' findings, including information or insights about the beginnings of tracking, the impact of tracking decisions, the role data plays in making tracking decisions, and outcomes associated with tracking students.

Tracking in Middle Grades Mathematics

In the aforementioned large school district in North Carolina, school counselors wanted to use students' achievement data and scholarly research to align services and opportunities better for students. As they started using achievement data and research findings, they became aware of the role mathematics learning plays in students' overall success in school. In particular, they discovered that performance in algebra affected graduation and dropout rates. Counselors also realized that they needed to examine assumptions about how they tracked students in mathematics. For example, many school counselors assumed that students who had scored at or above grade level on the North Carolina end-of-grade (EOG) assessments automatically tracked into the top middle school math classes. However, when they examined students' data, they saw that this was hardly true. Consequently, school counselors realized that collaboration among math teachers, school administrators, and themselves was important for addressing the effects of tracking associated with closing achievement gaps, offering more rigorous coursework for all students, and increasing graduation rates.

When Does Math Tracking Begin?

In this particular North Carolina school system, students leave elementary school in grade 5 to begin middle school in grade 6. From kindergarten to grade 5, students take the same math courses. On entering middle school, their math experiences begin to diverge because of tracking. Most students tracked into Math 6 or Advanced Math 6. Very few students skipped sixth-grade math altogether and took prealgebra or algebra. Interestingly, Math 6 and Advanced Math 6 used the same curricular materials. The districts' math educators typically described the difference between Math 6 and Advanced Math 6 like this: Advanced Math 6 is designed to prepare students for the most rigorous math sequences in middle grades and high school. It teaches for conceptual understanding and engages students in using higher-order thinking skills. Although rigor exists in Math 6, expectations for students are lower, because the district does not expect these students to take the more advanced math coursework in middle school or high school.

Once students were tracked low, no evidence showed that any student got the opportunity to move into the higher track. For every school in the district, the negative effects of the initial tracking decisions continued beyond grade 6. That is, if students were tracked low in grade 6, then they would be tracked low in grade 7 unless individual teachers recommended changes. However, no placement policy existed that would indicate how students might improve their standing in the tracking scheme. Hence, depending on the middle school, 0 to 2 percent of students placed in Math 6 took prealgebra in grade 7, whereas 70 to 90 percent of students placed in Advanced Math 6 routinely took prealgebra in grade 7. Math6 students typically took Math 7—a continuation of Math 6—with the same expectations from teachers.

What Is the Impact of Tracking Decisions?

No districtwide criteria existed for tracking students into sixth-grade math. Each elementary school used its own criteria—some formal, some not—to make placement decisions. Near the end of grade 5, all students take a state standardized EOG assessment in math. A student can score at Level 1, 2, 3, or 4. Students who score Levels 1 or 2 are said to be below grade level; Level 3, at grade level; and Level 4, above grade level. However, although all grade 5 students take the math EOG, fifth-grade teachers must make their sixth-grade math recommendations before the test results are available.

Eighth-grade algebra is arguably the gateway course to rigorous high school science and math. Prealgebra in grade 7 is the prerequisite for eighth-grade algebra. If students are tracked into Math 6, they probably will not gain access to rigorous high school courses. Hence, the key to rigorous coursework in high school is the sixth-grade placement in math.

To try to understand which students would probably be successful in eighth-grade algebra, we looked at the grade 5 EOG scores in math of the students enrolled in algebra in grade 8. Most students in algebra in grade 8 had, indeed, scored Level 4 on their grade 5 EOG math test. However, many students who scored Level 4 on their grade 5 EOG math assessment did not take algebra in grade 8. In fact, the likelihood that a student scoring Level 4 on the grade 5 EOG math assessment would be tracked high in middle school and go on to algebra in grade 8 differed greatly by the middle school attended, ranging from 20 percent to nearly 80 percent. Overall, fewer than half the students who scored Level 4 on their grade 5 EOG math assessment took algebra in grade 8. Furthermore, the percents of Level 4 students who were tracked high in math and went on to take algebra in Grade 8 differed by race and were statistically significant at a p value of .001 (see fig. 6.1).

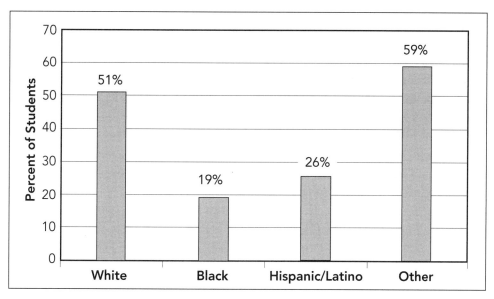

Fig. 6.1. Districtwide percent of grade 5, level 4 students who were tracked high in middle school and went on to grade 8 algebra

Examining the Impact of Tracking

No districtwide policy existed for assigning students to middle school mathematics. However, if such a policy did exist, the question would arise on whether the district should base those assignments on meeting objective criteria or on attitudes and beliefs about students and "what we know for sure." Seven middle schools out of thirty in the district agreed to participate in a study to determine the effects of placing all grade 6 students into a uniformly rigorous, sixth-grade mathematics course called Algebraic Thinking. In essence, they would have only one mathematics track, that of advanced math.

The school district's Curriculum and Instruction department supported the seven schools' participation in the study. The school system wanted to start with a small group of schools, using them as a pilot to examine teaching Algebraic Thinking. The Curriculum and Instruction department provided professional development and instructional support.

Since many teachers had not previously taught in the advanced track, the Curriculum and Instruction department instituted professional development designed to help teachers differentiate their instruction for a wide range of students' abilities, add scaffolding to lessons to support learning, and create rigorous lessons with the goal of teaching for conceptual understanding. The department arranged planning periods so that teachers could collaborate. They often made gifted- and special-education teachers available to help classroom teachers achieve their instructional goals by bringing in more professional development and appropriate curricular materials. A few of the middle schools had an Advancement via Individual Determination program, which is designed to help minority and low-income students succeed in rigorous classes.

At the end of the year, after students had completed a year of Algebraic Thinking, we wanted to see how their sixth-grade math teachers would place them in math in grade 7. We wanted to compare their academic success, as measured by standardized math tests, to that of students from the nonparticipating middle schools who, purportedly, were tracked homogenously by ability. We were also interested in seeing how grade 6 teachers' math placement decisions compared to those of grade 5 teachers. Students from the seven participating middle schools were the study group; the districts' remaining middle schools composed the control group.

How Is Data Used to Track Students?

For the seven middle schools that participated in examining tracking outcomes, grade 5 teachers completed paperwork as usual, recommending students for their sixth-grade math placement, not knowing that all students would take Algebraic Thinking regardless. As indicated above, teachers made their recommendations before standardized EOG math test scores were available. When EOG math scores became available, we compared teachers' predictions of students' EOG math scores to students' actual EOG math scores, by school and by race.

The school district required parents to read and sign the recommendation forms, acknowledging their agreement with the placement decisions. If a parent had concerns, he or she could write comments on the recommendation form. The parent could even sign a waiver that would override the teacher's recommendation if the parent believed that the child should be placed differently.

Fifth-grade teachers from 85 elementary schools completed course placement recommendations for students in the study group. Some of the study group schools were magnet schools and served students from many different elementary schools. We created electronic files recounting information from the placement recommendation forms. That information included teachers' predictions of students' standardized math test scores, math placement recommendations, and an indication of whether parents had signed waivers asking that their child be placed higher than recommended. We compared the growth (i.e., the improvement in test scores) on the EOG math assessments of sixth-grade students attending the seven middle schools in the study to that of similar students not in the study, by race and achievement level. We also compared course enrollment patterns of Level 4 students enrolled in Advanced Math 6 to that of Level 4 students enrolled in Math 6.

We contrasted grade 5 teachers' recommendations for students in the study group to those from grade 6 teachers for the same students the following year. We were especially interested in comparing the students who scored high yet received low-track recommendations from grade 5 teachers. After comparing teachers' recommendations, we surveyed and interviewed school counselors and teachers about what they believed to be true regarding the math placements and recommendations they made. We also surveyed and interviewed grade 5 teachers to determine what criteria they used in making placement recommendations.

Informal focus groups and a Web-based survey were used to determine what school counselors believed about the process of middle school math placements and what assumptions counselors made about students during the process. We also surveyed high school counselors and deans of student services about whether official or unofficial school or district policies existed regarding course enrollment related to middle school math placements.

Outcomes Associated with Tracking

To determine significance for the comparisons we were making, we used repeated-measures analysis of variance at a p value of .001, because we wanted to observe and compare the same students over time.

Fifth-Grade Teacher Predictions Compared to Actual Scores

Comparing teachers' predictions to actual EOG test scores, we found that teachers correctly predicted students' performance for about 60 percent of their students. For the nearly 40 percent of students for whom teachers made incorrect predictions, teachers equally overestimated and underestimated students' performance. However, estimates were not equally correct across ethnic groups. Teachers incorrectly predicted lower scores than non-Asian minority students actually received at statistically significant greater rates than they did for Asian or white students. Table 6.1 gives the percent of correct predictions for each demographic group and illustrates that teachers more accurately predicted the scores of Asian and white students than those of non-Asian minorities.

Table 6.1
Predicted Performance of EOG Level 4 Students on Grade 5 EOG Mathematics Assessment

Ethnic Group	Total Number	Percent of Total Number Correctly Predicted
White	1427	65%
Non-Asian Minority	892	54%
Asian	125	70%
Total	2444	61%

Sixth-Grade Teacher Recommendations Compared to Fifth-Grade Teacher Recommendations

After students had completed a year of Algebraic Thinking, sixth-grade teachers made recommendations for the next math course students should take. Grade 6 teachers recommended that 15 percent more students be placed into the advanced track leading to algebra in grade 8 than the grade 5 teachers did. This included the 6 percent of students whom the teachers recommended go directly into algebra after grade 6.

Although grade 5 teachers' recommendations for Advanced Math 6 correlated highly with the students' ethnicities, grade 6 teachers' recommendations for grade 7 prealgebra did not. They recommended about the same percent of Level 4 students of each ethnic group for grade 7 prealgebra, the course that leads to algebra in grade 8 (see table 6.2). In many instances, grade 5 and grade 6 teachers did not recommend the same students for their respective advanced tracks. In fact, one-fourth of the students recommended for the low track by the fifth-grade teachers received high-track recommendations from the sixth-grade teachers. Similarly, one-fourth of the students that fifth-grade teachers recommended for the high track did *not* receive high-track recommendations from the sixth-grade teachers. Moreover, all the students that the fifth-grade teachers recommended for the low track, but who had EOG math scores higher than the mean of the students the fifth-grade teachers recommended for the high track, received high-track recommendations for entering grade 7.

Table 6.2
Percent of EOG Level 4 Students Recommended for the Next High-Track Math Course

Ethnic Group	Recommended to Advanced Math 6 by Fifth-Grade Teachers	Recommended to Grade 7 Prealgebra by Sixth-Grade Teachers
White	71%	68%
Non-Asian Minority	12%	61%
Asian	83%	70%

Study-Group Students Compared to Control-Group Students

We compared study-group students' performance to that of control-group students who achieved similar grade 5 EOG scores and were placed into Math 6. Regardless of their ability level, study-group students had greater growth on standardized EOG mathematics assessments than students in the control group (see table 6.3). Similarly, per their ethnic groups, students from the study group were compared to similarly scoring students from the control group who took Math 6. Regardless of their ability levels, students of every ethnic group in the study group demonstrated greater growth than their cohorts in the control group (see table 6.3). Growth from grade 5 to grade 6 was measured by comparing the mean C-scores on the EOG math assessment. C-scores are computed using a regression analysis that includes the scores of all North Carolina students, in which students' observed scores are compared to their predicted scores.

Table 6.3

Mean EOG Math C-Scores of Students in the Study Group vs. Control Group by Ethnic Group and Achievement Level

(Study Group N = 2450, Control Group N = 3152)

Ethnic Group	Control Group (CG) Mean C-Scores	S.E.	Study Group (SG) Mean C-Scores	S.E.	Difference Score (SG – CG)
White	0.11	0.10	0.52	0.02	0.41*
Non-Asian Minority	0.39	0.19	0.46	0.02	0.85*
Asian	0.15	0.03	0.48	0.03	0.33*
Achievement Level					
Level 1	0.06	0.17	0.41	0.03	0.47*
Level 2	0.81	0.16	0.58	0.03	1.39*
Level 3	0.22	0.02	0.73	0.03	0.51*
Level 4	0.11	0.03	0.45	0.03	0.34*

*$p < .001$

Do Waivers Matter in Tracking Decisions?

As previously noted, parents could have signed waivers to change their children's math placements in Grade 6. Nearly all the signed waivers were from parents whose children had scored above the mean scale score of students recommended for Advanced Math 6 and had scored Level 4 on the EOG math assessment for two consecutive years. Ninety-eight percent of these students were identified as academically gifted in math; their standardized math percentiles ranged from 93 to 99. And yet, these students were recommended for the Math 6, the lower-track math course.

Whether parents of these high-scoring students signed waivers differed by ethnic group (see table 6.4). The fact that no black parents signed waivers is significant. Research has shown (Mallery and Mallery 1999) that minority parents believe they have little influence with school staff regarding their children's education, and they are therefore less likely to try. However, this does not mean that they are any less concerned about their children's education. Although they may be less apt to attend PTA meetings or volunteer for field trip chaperone, many help their

Table 6.4

Percent of Parents of High-Scoring* Students Signing Waivers to Gain Access to Advanced Math 6

Ethnic Group	
White	44%
Non-Asian Minority	20%**
Asian	83%

*These were students whose scale scores were higher than the mean scale score of students who were recommended for the top math track.
**No black parents signed waivers.

children in other ways, such as allowing less chore time so children can do their homework or sacrificing so their children can attend better schools (Auerbach 2007).

Do Attitudes and Beliefs Play a Role in Tracking Decisions?

Twenty-three fifth-grade teachers were interviewed and 84 fifth-grade teachers took a Web-based survey to determine the criteria they used to make tracking decisions in math. Not surprisingly, criteria differed among schools and frequently differed among teachers at a given school. Criteria included teacher-predicted EOG math assessment scores, course grades, and beliefs about students' work habits. Moreover, some teachers actually reported that they placed students on the basis of whether the teachers believed the students would need advanced math later in life. In fact, these teachers didn't believe that many people really needed to learn advanced math to be successful later in life. We should note that (1) no uniform criteria existed in the district for making such decisions and (2) the school or district provided no professional development or formal guidance to help teachers make their recommendations.

Interactions with school counselors from all the elementary, middle, and high schools in the district revealed that many of them believed that students who scored Level 4 on the grade 5 EOG math assessment were tracked high in middle school math and eventually went on to take algebra in grade 8. Counselors even expressed confidence that all Level 4 black students were tracked high in middle school math because of the district's initiatives to get more minority students into advanced science and mathematics courses. However, such confidence did not match patterns in districtwide data, which showed that only 19 percent of Level 4 minority students had actually taken grade 8 algebra. One middle school actually placed fewer students in Advanced Math 6 than its teachers recommended for it. During follow-up interviews, teachers at that school explained that the school did not want too many students prepared for eighth-grade algebra because the school was not prepared to teach more sections of the course. Unfortunately, this same school also recommended fewer Level 4 non-Asian minority students than Level 4 white or Asian students for Advanced Math 6.

Do Tracking Decisions Really Matter?

We looked at the course enrollment patterns of students who scored Level 4 on the grade 5 EOG math assessment who went on to take grade 8 algebra. We compared them to Level 4 students who took algebra in the ninth grade a year later, this being the typical math class for students who were placed on the standard track. Although many educators might argue that eighth- and ninth-grade algebra are the same course, with the same rigor, we found that school counselors and some teachers assumed that all students who scored Level 4 on the grade 5 EOG math assessment went on to take grade 8 algebra. This clearly was not true, and this fact led to some surprising outcomes.

Using districtwide data for students for whom we had complete data in grades 5–12, we found that 3,201 students had scored Level 4 on the grade 5 EOG math assessment. Of these, 51 percent had taken algebra in middle school and 49 percent had taken ninth-grade algebra. We discovered that Level 4 students' enrollment in high school science courses related to the grade at which the students had taken algebra. Nearly all (94%) the Level 4 students who took algebra in middle school went on to take chemistry, whereas 66 percent of the Level 4 students who took ninth-grade algebra did. Fifteen percent of these Level 4 students who took algebra before ninth grade took physics, three times the number (5%) who took ninth-grade algebra.

Tracking decisions apparently do affect students' future opportunities and performance. Although many consequences of tracking capable students into the low math track are apparent (Burris, Heubert, and Levin 2006), other harmful results may go unnoticed. Indeed, some school counselors and teachers made unnecessary referrals for students to intervention programs. Many also recommended students for courses and interventions on the basis of an assumption that 100 percent of Level 4 minority students had taken algebra in middle school. But, as previously reported in figure 6.1, far fewer non-Asian minority students had actually taken grade 8 algebra.

For example, after the fifth-grade teachers made their recommendations for students' grade-6 placements, they often assigned minority students to the school district's dropout prevention program, having assumed that minority students who were *not* in the top math track in middle school must be low scoring in math. Minority and low-income students were viewed as at-risk students, and the dropout prevention programs were designed to help at-risk students improve math skills. Eventually, the school district learned that it had assigned a disproportionate number of Level 4 minority students to dropout prevention programs and excluded a disproportionate number from many rigorous math and science courses. Soon after, school counselors in the district began routinely using actual performance data for course enrollment and instructional intervention decisions.

What We Know for Sure

This study revealed that the beliefs, attitudes, and prejudices of school counselors, math teachers, and administrators often override much more objective information. The partnership of counselors, math teachers, and administrators with the Data Academy facilitators opened the door to using objective criteria for making academic decisions about students, especially decisions about tracking in mathematics. Although we shared the study's findings of the study with school personnel and school counselors have generally accepted them, math teachers remain reluctant to accept what the data show about students' performance and proper placement of students, often citing students' behavior—and weaknesses in the EOG assessments themselves—as reasons for ignoring recommendations based on objective criteria. Since administrators tend to follow mathematics teachers' lead rather than that of school counselors regarding math tracking, resistance remains strong to making tracking decisions using objective criteria.

Although some schools embraced the findings of the study and implemented using objective criteria for placing students, many more have not. It seems that what we know for sure often blinds us to the realities shown in data. This is unfortunate, given that the set of objective criteria were established jointly by the departments of the Academically Gifted, School Counseling, and Curriculum and Instruction, on the basis of (1) what the data showed about future success in mathematics and (2) the sensibilities of school personnel. In fact, objective criteria were established for acceptance into all the rigorous math courses in middle school. In particular, to be placed in Advanced Math 6, students had to meet two of three criteria: (1) having a supportive teacher recommendation, (2) scoring a high Level 3 or a Level 4 on the EOG assessment, and

(3) receiving a grade no lower than a 3 out of 4 on their standards-based report cards and two work samples. If students met two of the three criteria, school counselors would recommend placement in Advanced Math 6. However, since school board policy does not govern such recommendations, the recommendations may be ignored.

Ignoring such recommendations has unfortunate results, especially for students moving into the middle grades. As noted elsewhere in this chapter, access to high-quality mathematics content and instruction in the middle grades is essential to students' future opportunities and success. Clearly, establishing a set of criteria is not enough; school board policy must mandate using the objective criteria districtwide.

Using data to make important decisions about students' performance can prevent unintended negative consequences, such as the improper labeling of students and the concomitant, inappropriate teaching and learning interventions. Interestingly, school personnel do not question placing a student labeled academically gifted into the next rigorous mathematics course, whether or not the student meets two of the three objective placement criteria.

Our work suggests that all students need access to rigorous, high-quality mathematics and that students tend to perform at the level we establish for them, be it high or low. The research (Hallihan 2003; Garrity 2004; Burris, Heubert, and Levin 2004) has demonstrated this and cautions us to rely on properly collected data to determine what we know for sure. In fact, on the basis of our work, we would recommend that (1) the practice of tracking in mathematics be eliminated in grade 6, giving all students equal access to the best math content and instruction available in a school; (2) school personnel, especially school counselors, work with parents and students to help them understand school policies and practices that may adversely affect them; and (3) school personnel, especially school counselors, make students' performance data available to parents and students on a regular basis and explain students' options for future success.

Although all students should have access to the best math courses we have to offer, it is especially disappointing that many math teachers refuse to grant access to students who have already demonstrated an ability to achieve in mathematics. "What we know for sure" about low-income and minority students' capacity to learn mathematics has blinded us to the realities of what the data say about students' past and future performance. Hopefully, educators will heed what Mark Twain has correctly noted, "It ain't what we don't know that gets us into trouble. It's what we know for sure that just ain't so."

REFERENCES

Akos, Patrick, Marie Shoffner, and Mark Ellis. "Mathematics Placement and the Transition to Middle School." *Professional School Counseling* 10 (February 2007): 238–44.

Auerbach, Susan. "From Moral Supporters to Struggling Advocates: Reconceptualizing Parent Roles in Education through the Experience of Working-Class Families of Color." *Urban Education* 42, no. 3 (2007): 250–83.

Burris, Carol Corbett, Jay P. Heubert, and Henry M. Levin. "Accelerating Mathematics Achievement Using Heterogeneous Grouping." *American Educational Research Journal* 43 (Spring 2006): 105–36.

Callahan, Rebecca M. "Tracking and High School English Learners: Limiting Opportunity to Learn." *American Educational Research Journal* 42 (Summer 2005): 305–28.

Choy, Susan P. *Access and Persistence: Findings from 10 Years of Longitudinal Research on Students.* Washington, D.C.: American Council on Education, Center for Policy Analysis, 2002.

Garrity, Delia. "Detracking with Vigilance: By Opening the High-Level Doors to All, Rockville Centre Closes the Gap in Achievement and Diplomas." *School Administrator* (August 2004). http://findarticles.com/p/articles/mi_m0JSD/is_7_61/ai_n6172415/, December 21, 2009.

Gray, Jeremy R., Todd S. Braver, and Marcus E. Raichle. "Integration of Emotion and Cognition in the Lateral Prefrontal Cortex." *Proceedings of the National Academy of Sciences* 99 (March 2002): 4115–20.

Hallinan, Maureen T. "Ability Grouping and Student Learning." *Brookings Papers on Education Policy 2003* (2003): 95–124. http://muse.jhu.edu/journals/brookings_papers_on_education_policy/v2003/2003.1hallinan.html.

Hopkins, Gary. "Is Ability Grouping the Way to Go—or Should It Go Away?" *Education World*, February 24, 2009. http://www.educationworld.com/a_admin/admin/admin009.shtml.

Johnson, Janet L., Bernice Campbell, Rita Lewis, Julie Johnson, Marty Redington, and Aniko Gaal. *North Carolina's School Counseling Program Review: A Statewide Survey and Comprehensive Assessment.* Vol. 2006. Raleigh, N.C.: EDSTAR, Inc., 2005.

Mallery, James L., and Janet G. Mallery. "The American Legacy of Ability Grouping: Tracking Reconsidered." *Multicultural Education* 7, no. 1 (Fall 1999): 13–15.

Stone, Carolyn B., and Robert Turba. "School Counselors Using Technology for Advocacy." *Journal of Technology in Counseling* 1 (October 1999). http://jtc.colstate.edu/vol1_1/advocacy.htm.

Vanfossen, Beth E., James D. Jones, and Joan Z. Spade. "Curriculum Tracking and Status Maintenance." *Sociology of Education* 60 (April 1987): 104–22.

Wheelock, Anne. *Crossing the Tracks: How "Untracking" Can Save America's Schools.* New York: New Press, 1992.

Mathematics Education, Language Policy, and English Language Learners

Marta Civil

S INCE 2004, researchers in the Center for the Mathematics Education of Latinos/as (CEMELA) have been studying issues related to the interplay of culture, language, and mathematics teaching and learning with Latino and Latina students. One of the topics of study is how different language policies may affect students' participation in mathematics education, as well as that of their parents. Acosta-Iriqui et al. (in press) address the impact of two very different language policies, one in Arizona and one in New Mexico, on Latina mothers' engagement in their children's learning of mathematics. The New Mexico Constitution endorses bilingual education. Arizona, however, severely restricted bilingual education in 2000 by passing Proposition 203, inspired by California's Proposition 227, passed in 1998. Since Proposition 203's passing, Arizona has required its public schools to place English language learners (ELLs) in Structured English Immersion (SEI) classrooms, where instruction is only in English and teachers are allowed to use a student's first language only for clarification (Combs et al. 2005).

An assumption behind Proposition 203 was that after one year of SEI, students would have acquired enough English to be academically successful. Experts in second-language acquisition do not support this assumption, indicating instead that acquiring academic English takes much longer than one year (e.g., Cummins [2000]). The linguistic consequences behind this law's passing need to be interpreted from a political viewpoint as well as an opportunity-to-learn consideration. For example, one possible educational outcome associated with restrictive language policy is *subtractive schooling* (Valenzuela 1999), which strongly discourages and often prohibits ELLs from using their first language, thus risking its loss.

Moreover, in the public discourse—as reflected, for example, in comments by the public in general in newspapers, blogs, and so on—it is hard to separate a proposition that apparently addresses only the language of instruction from the political ramifications of whose opportu-

The Center for the Mathematics Education of Latinos/as (CEMELA) is a center for learning and teaching funded by National Science Foundation grant ESI-0424983. The views expressed in this chapter are those of the author and do not necessarily reflect those of the funding agency.

nity to learn it affects the most. As Varley Gutiérrez (2009) writes, "[A]lthough this legislation [Proposition 203] is focused on restrictive *language* policies, it reflects a national trend at the time of this study, toward policies and practices marginalizing the growing Latina/o population in the United States" (p. 95). Wright's (2005) analysis of Proposition 203 argues for a political motivation behind this initiative rather than a real concern for the education of ELLs. The discussion needs to refocus on matters of opportunity to learn, rather than on ideological debates about language.

This chapter focuses on the intersection of research, practice, and policy in teaching mathematics to, and the mathematics learning of, Latino ELL students in Arizona. Mathematics educators, including this chapter's author, have seen the effect of a language policy such as Proposition 203. The chapter will briefly describe cases that illustrate how research, school practice, and language policy interacted as sites of discovery to guide CEMELA discussions of opportunity to learn mathematics. The cases follow three themes: placement decisions in mathematics classes; language, mathematics, and subtractive schooling (Valenzuela 1999); and the complexity of language ideology in the mathematics classroom. Before discussing the cases, the chapter will examine some background on the history of CEMELA. Specifically, it will focus on the research framework driving the center's theory of action and the structure of the partnerships that support programmatic efforts. This background information is relevant because the cases are outcomes of CEMELA's theory of action.

Situating the Work: A Brief Overview of CEMELA

CEMELA is a center for learning and teaching, started in 2004 and funded by the National Science Foundation. Its primary mission is studying the interplay of mathematics education and the unique language, social, and political issues that affect Latino communities. A consortium of four universities in partnership with several school districts, CEMELA carries out research and practice in four different settings—borderlands, migrant agricultural, rural, and urban. The focus is primarily on Mexican and Mexican American communities. CEMELA grounds its research in a sociocultural perspective with a particular emphasis on community knowledge (Civil 2007; González, Moll, and Amanti 2005) and language (Khisty and Chval 2002; Moschkovich 2002, 2007).

A primary concept in CEMELA's research is *funds of knowledge* (González, Moll, and Amanti 2005), which underscores the linguistic and cultural resources in Latino communities that can support children's learning in school. CEMELA approaches the mathematics education of Latino and Latina students holistically, through a research agenda based on working with teachers, parents, and students. Thus, CEMELA's research studies are action-based in that they (1) work collaboratively with the interested parties toward developing mathematically rich learning environments and (2) build on the experiences and backgrounds of children and families.

CEMELA's work with teachers and parents combines explorations of mathematics with discussions about teaching mathematics to, and the mathematics learning of, Latino and Latina children (Civil 2009) with an emphasis on challenges and affordances. CEMELA's work with students takes place in two settings—classrooms and after-school mathematics clubs. Turner,

Celedón-Pattichis, and Marshall's (2008) study of three kindergarten classrooms underscores the importance of using classroom practices that build on children's cultural and linguistic resources. Their study is a powerful example of opportunity to learn (Tate 2005), because we see students participating in mathematical discourse and using multiple strategies to solve complex arithmetic problems. This study took place in bilingual classrooms where the use of students' home language, Spanish, was encouraged and supported. In contrast, Civil (2008c, 2009) and Planas and Civil (2009) address the challenges to participation in mathematical discourse in classrooms where the use of one's home language, by law, is not supported. This chapter will illustrate some of these challenges.

CEMELA studies in the after-school mathematics clubs highlight some of the tensions around community knowledge and language choice—English or Spanish. Despite the tension, when doing mathematics we made an effort to promote of the use of both Spanish and community contexts (Khisty and Willey 2008; Turner et al. 2009). These tensions point to the deep-rooted effects of a schooling process that devalues Latino children's language and culture. This is particularly true when restrictive language policies are in place. CEMELA has developed a research program that examines language policy and opportunity to learn mathematics. A discussion of this effort follows.

Placement in Mathematics Classes

Civil (2008b) presents a survey of the mathematics teaching and learning of immigrant students in different parts of the world. A point in common among mathematics teachers in many European countries, the United States, and Australia seems to be their view of language as a problem (i.e., students not knowing the language of instruction well), often conveying a deficit view of second-language learners. This contrasts with the call by researchers in mathematics education in multilingual contexts for the need to capitalize on the resources that multilingual students bring to the classroom (Moschkovich 2002). A focus on "language as an obstacle that has to be overcome" may lead to placement decisions that have implications for ELLs' academic advancement. Placing students in lower-level courses with the idea that first they need to learn the language is a practice that has serious equity implications. In her recommendations for the education of immigrant students, Valdés (2001) argues for the need for students to have access to the mainstream curriculum while they are learning English. As she writes, "[S]tudents should not be allowed to fall behind in subject-matter areas (e.g., mathematics, science) while they are learning English" (p. 153). Similarly, Callahan, Wilkinson, and Muller (2008, p. 181) write,

> ESL [English as a Second Language] coursework is neither meant to replace, nor preclude access to, rigorous academic coursework; if it does, then it may seriously disrupt long-term academic trajectories. Mexican-origin students' access to academic content via course placement merits careful consideration.

This chapter will describe two different scenarios from our local context that point to a possible problem with how ELL students are sometimes placed into mathematics classes. Part of CEMELA's professional development work with teachers offered a series of mathematics content courses for middle school teachers from schools with large numbers of Latino and Latina students. For one session in one of those courses, a Chinese mathematics instructor conducted a lesson on area and perimeter in Chinese. The goal was to have the teachers experience being second-language learners in a mathematics class (for details, see Anhalt, Ondrus, and Horak [2007]). During the debriefing, one teacher shared how she had basically ignored the language and focused on the mathematics because she was familiar with the content. This made her question her school's placement policy, which had placed some ELL students in mathematics classes below their perceived math skills and understandings (Anhalt, Ondrus, and Horak 2007, p. 22):

> The thinking behind this placement policy is that the students will learn English through the mathematics content they already know. [The teacher] pointed out that based on her experience during the Chinese lesson, these students could very possibly be ignoring the English language completely during instruction, just as she ignored the Chinese language. Further, she pointed out that the students might be bored because the lower-level mathematics was too easy for them. She wondered if ... school systems are doing students a disservice in both their English language development and mathematics learning.

The second scenario is from CEMELA's work with parents. One of the goals is understanding how immigrant parents perceive their children's mathematics education, and their schooling in general, in the United States. One recurrent theme in the data is parents' perception that the mathematics level is higher in Mexico than in their children's current U.S. school (Civil and Planas 2010). This was true of Emilia and her son, Alberto: they both commented that what he was studying in U.S. sixth grade, he had already studied before in Mexico (Civil 2008a). In March 2006, the mother said,

> What I feel is that they teach them more things there [in Mexico]. Now the difference here is that you have the language, ... and so for them it's perfect what they are teaching them because in this way it's going to help them grasp it, to get to the level, because for them, with the lack in English that they have, ... if they give them all the information... too much teaching during this period, to tell you the truth, it would disorient them more. Right now, what he is learning, what I see is that it's things that he had already seen, but if he gets stuck, it's because of the language, but he doesn't get stuck because of lack of knowledge.

Emilia seems to be fine with the fact that her child may be seeing content that he already knows, because she thinks that the priority at this stage is for her son to learn English. Later in the interview she makes a very insightful observation:

> What I see is that on their homework assignment it says "grades 6–8." What I gather from this is that they are getting, I mean, that a child in grade 7 and one in grade 8, they are doing the same thing?

We interviewed Emilia again in December 2007; by then Alberto was in eighth grade and Carlos, her middle son, was in seventh grade. They both had the same teacher for mathematics, a teacher specifically assigned to the ELLs. Emilia said,

> What I sometimes don't understand is why they give them the same homework if they are in different grades.... It bothers me a bit because it leads me to believe that, as if the eighth is at the same level as the seventh, you know? One assumes that the eighth is at a higher level.

Emilia captures quite well what CEMELA thinks is problematic about how ELLs are placed in mathematics classes. A focus on students' need to learn the language of instruction may be hampering their advancement in academic content areas such as mathematics. The implications of school language policies for students' placement in courses need to be examined and even confronted, because they may be obstacles to an equitable mathematics education (NCTM 2008).

Language, Mathematics, and Subtractive Schooling

This section will argue that restrictive language policy contributes to a classroom experience that inhibits growth in both mathematics learning and language development. CEMELA's research indicates that restrictive language policy is foundational to an educational experience that Valenzuela (1999) refers to as *subtractive schooling*. Giving a thorough account of Valenzuela's work on subtractive schooling is not possible. A central message from her work with immigrant Mexican and Mexican American students, however, is that for most of the high school students in her study, "schooling is a *subtractive* process. It divests these youth of important social and cultural resources, leaving them progressively vulnerable to academic failure" (p. 3). Valenzuela's ethnographic study of Seguín High School describes the consequences of this negation of cultural, social, and linguistic resources clearly (p. 62):

> Rather than building on students' cultural, linguistic, and community-based knowledge, schools like Seguín typically subtract these resources. Psychic and emotional withdrawal from schooling are symptomatic of students' rejection of subtractive schooling and a curriculum they perceive as uninteresting, irrelevant, and test-driven.

Moll and Ruiz (2002) also use Valenzuela's (1999) concept of subtractive schooling and note how it is "a major feature of the education of poor and working-class Latino students all over the country. It results in disdain for what one knows and what one is, influences children's attitudes towards knowledge, and undermines their personal competence" (p. 365). CEMELA's work looks at how these subtractive schooling practices affect parents' and students' participation in mathematics teaching. What follows will provide one example of each.

Acosta-Iriqui et al. (in press) illustrate the impact that different language policies have on parental participation in their children's mathematics education, particularly when the parents do not know English well. In our local context, evidence shows how the changes from bilingual education to English-only affected parents' support of their children with homework and with visits to their children's classrooms. More specifically, some parents stopped going to classrooms because they could not understand the instruction (Civil 2008a; Civil and Planas 2010). This is happening at the same time when schools feel pressure to increase parental involvement from legislation such as the No Child Left Behind Act. Although a description elsewhere has argued the need to expand the definition of parental engagement beyond parents' physical presence in schools (Civil and Andrade 2003), a concern persists that these restrictive language policies may alienate Latino, Spanish-speaking parents even more from schools. This alienation, in turn, feeds a common misperception among teachers and school administrators that "these parents don't care about their children's education." All the parents we have worked with *do* care immensely about their children's mathematics education and schooling in general. Some come to the mathematics workshops we have offered to bridge that language gap and allow them to help their child. For example, Camelia and Marcos, the parents of a seventh grader, came to learn the mathematics in Spanish so that they could better support their daughter, even though the language difference was still an issue. In an April 2007 interview, they said the following:

Camelia: If we work with her, here [at the school] they teach them in English, and when the girl is doing homework, we sit with her and it is in English, how can we help? [*raises her hands as if to indicate her frustration*]

Marcos: It's in English. The problem is in English.

Camelia: But not here [at the workshops]. Here we come, and you give it to us in Spanish. And it's the same that they do here [at school]. It is the same in English and in Spanish, the same fraction. But how can we help her if she counts in English, or …?

Marcos: Because for her [the daughter], the numbers, she knows them more in English than in Spanish.

Camelia: That is, she knows the number in Spanish but since here they are not practicing them in Spanish, she …

Marcos: She forgets it.

This excerpt points to a particularly worrisome theme—the loss of the home language, or rather, the missed opportunity to learn not only English, but also academic Spanish. The home language in most of our cases is Spanish. Children such as Camelia and Marcos' daughter speak Spanish at home. Yet, the subtractive schooling practices (Valenzuela 1999) rob these children of the opportunity to learn two languages well, for social as well as academic communication. As Hernandez, Denton, and Macartney (2010, p. 22) write,

> The vast majority of children in immigrant families are American citizens because they were born here, yet their diverse languages and emerging bilingualism represent an extraordinary cultural resource as the U.S. seeks to position itself in the increasingly competitive and multilingual global economy.

Children in contexts where restrictive language policies send them messages that their home language is not valued are missing a chance to become this "extraordinary cultural resource" that Hernandez, Denton, and Macartney (2010) mention.

Subtractive schooling does not just influence language development negatively; its supportive principle, restrictive language policy, links to mathematics classroom practices and opportunities to learn. To illustrate further how language policy affects participation in the classroom, consider the case of Carolina, who arrived in the United States from Mexico when she was in fifth grade. She was placed with a teacher who spoke Spanish. Mrs. Baker, herself a Mexican immigrant as a child, was teaching in a fourth- and fifth-grade SEI classroom. We determined, from her teacher's report, what we saw in the classroom, and Carolina's self-report in an interview, that Carolina's level of mathematics was quite good. Carolina told us that she had already learned earlier in Mexico most of what she was taught in U.S. fifth grade.

Mrs. Baker told us the following in a November 2005 interview:

> Well, you know with Carolina, she is very smart. She picks up stuff right away. Those are things that I notice. Like she knows her facts, multiplication. And I told the kids "next week we are going to start off with division," because I haven't done that yet. And she and Delia [a student who had arrived from Mexico the year before] were the ones that said, "Oh, I understand, know my facts, I already know how to do the division." So hopefully the two can be models to the other ones.

Mrs. Baker shared with us, and we confirmed through classroom observations, that she relied on Delia to translate for Carolina. When working in small groups, Carolina could participate. She could use Spanish, and in many instances children in her group could follow it, since several children at that school are bilingual. But in whole-class discussion Carolina hardly ever said anything, because most of the instruction was in English. Mrs. Baker was certainly aware of and frustrated by this situation. She said, "I feel that I'm not giving her [Carolina] that attention, that instruction that I am supposed to be giving her because she is the only one who can't speak English."

Managing mathematical discourse in two languages is not an easy task; a culture in which using both languages is not the norm makes this task even harder. Mrs. Baker tried to involve Carolina in the whole-class discussion by translating into Spanish and inviting her to participate. Mrs. Baker, however, like other bilingual teachers we observed, had to make a conscious effort to remember to translate. Because the current language policy discourages using Spanish, whole-class discussions quite often are only in English. During an interview, Mrs. Baker expressed frustration at being underused, a bilingual teacher working in an educational setting where policy severely restricts her skills and understandings as a bilingual expert. An environment that does not nurture both—or all—languages present can make even small-group communication problematic.

In the episode described below, Carolina, the fifth grader; Griselda, a fourth grader and Carolina's sister; Darla, a fourth grader who speaks only English; and Anthony, a bilingual fifth grader, are working on the question "What do you notice about the fractions 1/2, 1/3, 1/4, 1/5?" This question was preparation for a task where students had to determine whether those fractions would split a group of 12 balloons and then a group of 12 brownies. A chart on the board listed the fractions 1/2, 1/3, 1/4, 1/5, 1/6, 1/8, 1/10, and 1/12, and labeled two blank columns beside the list—one, balloons; the other, brownies. The teacher approached this group of four students.

1. Mrs. Baker: What do you notice about the fractions 1/2, 1/3, 1/4, 1/5? Look at the board.

2. Darla: But what if they [*referring to Carolina and Griselda*] want to say something, and we don't understand.

3. Mrs. Baker: Well, he [Anthony] can translate for you.

4. Darla: OK.

Civil (2008c) presents more details on this case and, in particular, points out how difficult it was for Anthony to translate in a mathematical context. Once again, we see the missed opportunity to develop students' academic Spanish. Through the first part of this group's interaction, Griselda is dominating the conversation, but what she is trying to say is not clear from a mathematical point of view. Her sister, Carolina, finally intervenes and explains that the fractions, from 1/2 to 1/12, are getting smaller.

1. Carolina: Un medio. Cada vez se va haciendo más chico. [One half. Each time it's getting smaller.]

2. Griselda: Se está haciendo más grande. [It is getting bigger.]

3. Carolina: Griselda, estoy hablando. No digas más grande. Se hace más chiquito porque un medio es más grande y un tercio es más chiquito. El número, el numerador de abajo quiere decir que cada [vez] se va haciendo más y más… ¿cómo se dice … más …? [Griselda, I'm talking. Don't say it's bigger. It gets smaller because one half is bigger and one third is much smaller. The number, the numerator on the bottom means that each [time it] gets more and more … how do you say … more …?]

4. Anthony: ¿Largo? [Bigger?]

5. Carolina: No, más chico. Pues si es un [*pause*] doce, es algo muy chiquito, como así (*gesturing with her hands, showing a small separation*) y un medio es algo grandote. El numerador más bajo es más grande y el numerador más grande es más … más poquito. Menos cantidad. [No, smaller. So if it's one twelve, it's something really small, like this (*gesturing with her hands*) and one half is huge. The lowest numerator is bigger and the largest numerator is more … it is smaller. Less quantity.]

6. Darla: I have to learn Spanish.

This episode demonstrates several issues. Carolina seems to understand that as the denominator gets larger, the fraction gets smaller, but she first uses *number* (número, line 3 in the immediately previous dialog), and then switches to *numerator* (numerador), which she keeps for the rest of the dialogue, although she refers to the *numerator on the bottom* (el numerador de abajo). When she cannot think of the Spanish word for *smaller,* Anthony (line 4) provides *largo* (bigger), which makes us wonder what he understands, both mathematically and linguistically, from what Carolina just said. Carolina's gestures and sentences (line 5), although incorrect in her use of *numerator,* point to an understanding of what is going on with the fractions. Anthony does not offer any useful translation in this episode, which leads to Darla's comment in line 6, "I have to learn Spanish."

What did each student in this group learn about mathematics as he or she worked on this task? Carolina seemed to have a good command of fractions. Did she have an opportunity to share her knowledge with the whole class? What do teachers need to know about students like Carolina and Griselda to encourage their participation in the mathematics classroom?

Overall, most teachers with whom CEMELA has worked use group work in their approach to mathematics instruction. In many instances, they also used standards-based curricula (e.g., Investigations, Connected Mathematics). Students thus had an opportunity to engage in mathematical communication. In their small groups they often used both languages, English and Spanish. The issue remains, however, of how to engage ELLs in a contradictory environment. On the one hand, their home language is very present—at home, in the community, and even in the school. On the other hand, state policy sends a clear message that English is the language to be used in academic contexts, and teachers are reluctant to use a language other than English for fear of being reprimanded. Monitoring small-group discussions for mathematical learning and accuracy is not easy, because the teacher needs to distribute her time among the different groups. In most instances, the only language in the whole-class discussion was English, which did not allow us to see what would have been said if both languages were used.

The next and final section focuses on one experience in which we often used both languages, in small groups *and* with the whole class. Although this may have seemed an ideal situation, the segregated environment in which it happened underscores the complexity of language ideology.

The Complexity of Language Ideology in a Mathematics Classroom

In 2006, to further compound the restrictive effects of Proposition 203, the Arizona state legislature passed H.B. 2064, a new law that increased state funding to public school districts that serve English language learners. In order to qualify for the funds, however, the law required that districts place their ELLs in a segregated, four-hour, daily English block on grammar, reading, vocabulary, and writing, separate from their non-ELL peers. As of this writing, if the districts in this study wanted the state funding that the law offered, they have had no choice but to implement this model.

During the 2007–08 academic year, one of the middle schools where CEMELA conducted research initiated a program very similar to H.B. 2064's four-hour model. All the school's ELL students had five to six of their seven classes together in a specific section, hereafter called Section A. The program was designed to place students in smaller groups where teachers could, in theory, better address their needs as language learners.

The experience was reminiscent of Valdés's (2001) two schools in one building. From November 2007 to the end of the school year in May 2008, the author worked regularly with a seventh-grade class that had only eight ELL students, seven boys and a girl. Seven of the students were recent immigrants from Mexico, and most of them had arrived as fifth graders. The mathematics teacher was also from Mexico and an English language learner. The teacher taught mostly in English and used Spanish for clarification, usually in the small groups. The books and any handouts were in English, although the teacher had one set of textbooks in Spanish.

From a research perspective, I found the situation somewhat puzzling, because I wanted to focus on these students' understanding of mathematics and the language issue was a limitation. We were all native speakers of Spanish using English to communicate. I tended to use both languages, sometimes saying something in one language and repeating it in the other. As time went on in this class, the teacher and I developed the classroom norm that students were expected to explain their work. We were constantly asking students to justify their claims. Although most of the work was in small groups, we also asked students to present to the whole class from time to time. We let the students use whatever language they wanted, and we—I more than the teacher—did the same thing.

One of the first times that we asked groups to present their work to the whole class happened in March 2008. The students had been working on one of the activities in the Filling and Wrapping unit (Lappan et al. 2006). They had made triangular, square, pentagonal, and hexagonal prisms and a cylinder, each using one sheet of paper. The task was to find the volume of each of these objects by filling them with two-centimeter cubes and then comparing their volumes. One pair of students, Carlos and Larissa, had actually figured out the volume in cubic centimeters by multiplying the number of cubes by eight. Thus their answers—and most important, their approach—were quite different from those of the other students who had counted the number of cubes and recorded that number as the volume. We asked these two students to present to the whole class.

1. Larissa: First, then we (pause) make the base and measured with how many cubes are ...?

2. Carlos: Cubes.

3. Larissa: Cubes are could be ..., be filled ..., filled.

4. Carlos: Filled.

5. Larissa: Yeah. Sigue tú, ya, sigue tú. [Yeah. You go on, go on.] (*Looks down*)

6. Carlos: And then, ¿luego qué hicimos? [And then what did we do?] (*Carlos and Larissa whispering to each other*)

7. Larissa: Sacamos, sacamos el, sacamos el volumen. [We found, we found the volume.]

8. Carlos: We get the volume of centimeter and (*pause*) because one cube, an example, (*grabs and holds up a cube*) one cube, uh have eight centimeter.

9. Larissa: Eight centimeter ... of the volume.

10. Carlos: So, it's, like example, here are like fourteen cubes so we multiply uh, fourteen times eight and then you get .... (*Looks at Larissa*)

11. Larissa: How many cubes.

12. Carlos: How many ... the volume. (*Puts the cube back in the bucket*)

13. Carlos: Uh. ¿Lo agarraron? [Uh, did you get it?] (*Smiles*)

Carlos's and Larissa's nonverbal behavior while presenting is hard to convey in writing. They appear nervous—several instances of nervous laughter throughout, looking down at the floor, and so on. They look very tentative in their talk and whisper to each other. The key to their mathematical explanation is in lines 8–12, when Carlos takes a sample cube and says that it is eight centimeters. (It should have been eight cubic centimeters.) Then in line 10, he talks about multiplying the number of cubes by eight to get the volume. In line 13, Carlos asks the class if they "got it." Most students do not answer; some say yes. I did not think the explanation was clear enough, and said, "I don't think they got it." Carlos then asks if he can explain it in Spanish.

1. Carlos: ¿Lo puedo explicar en español? [May I explain it in Spanish?]

2. Carlos: ¿Para que lo agarren mejor? [So they can understand it better?]

3. Author: Yeah, you can say it in Spanish.

4. Carlos: ¿Sí? Mira, entonces por ejemplo aquí caben once cubos. (*Holds up cube and points to the base of the triangular prism*) [Yeah? Look, so for example eleven cubes fit here.]

5. Larissa: En la base. [In the base.]

6. Carlos: Uh-huh, aquí en la base no más. Y cada cubo en centímetros, mide ocho. [Yeah, here in just the base. And each cube in centimeters is eight.]

7. Larissa: El volumen. [The volume.]

8. Carlos: El volumen. Entonces multiplicamos, once cubos que había por ocho. Y ahí va a acabar el resultado, de centímetros … en cubos. ¿Ve? [The volume. Then we multiply, eleven cubes that were there by eight. And this is going to give us the result, in centimeters … in cubes. You see?]

Certainly mathematical issues exist in this excerpt (e.g., the issue of units on lines 6 and 8, cubic centimeters versus centimeters), as well as in the rest of their explanation (e.g., mixing up measurements in centimeters with number of cubes). But I want to point out the difference in demeanor between the explanation in English and the one in Spanish presented above. During the one in Spanish, both students appear confident and smiling. Although one could argue that technically they were saying the same thing in both excerpts, this was not the impression one got when watching them. In the episode in Spanish, they were more engaging and appeared more in command of the problem. After the excerpt shown here, I intervened and redirected them to use the whiteboard and record their work in cubes first. All along, I asked them questions in English, but in a very simple, direct dialogue (e.g., "How many cubes did you get for the height of the prism?"). They successfully reached the point where they had a volume of 121 cubes, and then they multiplied by eight to get an answer in cubic centimeters. At this point, one student said, in Spanish, "One question; I don't understand the eight. Where does the eight come from?" This initiated one of the best mathematical discussions we had that year. Five of the students—Larissa, Carlos, and three others—were particularly engaged, arguing in Spanish. I only intervened minimally at the beginning and again toward the end to see if a sixth student understood the discussion.

I learned a lot from this experience, as from others later in the semester. Students who appeared not to be engaged all of a sudden participated because they enjoyed arguing in Spanish, as they said in later interviews. Thus, they could transfer their experiences with arguing to the mathematical classroom. Being able to do that in Spanish played a pivotal role, showing us students' mathematical thinking in ways could not have happened had we limited ourselves only to English. We could also push these students mathematically in ways that did not go unnoticed. For example, Carlos noted in an interview that in other classes, when they were done with their work, they could go play on the computers, whereas in this class we always had them do mathematics and gave them more problems if they finished earlier. Carlos, Larissa, and another student from this class were placed in Algebra 1 in eighth grade.

Although our data supports the claim that these students engaged in mathematically rich discussions, and we believe that our ability to engage with them in Spanish and to let them use Spanish had a lot to do with this richness, the situation is much more complex than letting them use both languages. Interviews with the students—and in some instances, with parents—showed that these adolescents were very aware of their segregation in Section A (see Civil and Menéndez [2010]; Planas and Civil [2009]). Most of them expressed a desire to move out of Section A, and some believed that they were not learning as much English as they would if they were with the non-ELL students. Thus, in retrospect, it is not entirely clear that these students were necessarily comfortable with the idea of using Spanish in the mathematics classroom, since that may have contributed to their perception that they were not advancing enough in their English.

Closing Thoughts

This chapter underscores the complexity of the situation. If ELLs are in a class where English is the only language of instruction, the teacher may lose opportunities to gain more robust evidence of these students' mathematical understanding, as well as of areas where they need support. Insisting on English-only may limit their participation in small-group discussions, if they participate in class at all. Although segregating ELLs may increase their participation and allow teachers to learn more about the students' mathematical understanding by using more Spanish, the practice does not address all the students' needs, particularly the students' feelings about the segregation itself. Behind this oversight lie the messages sent by restrictive language policies such as Proposition 203, which make ELLs think that they are worth less because they do not know English well. Limited placement, limited participation, and segregation will not offer these students an equitable mathematics education.

This chapter highlights the need for a dialogue among policymakers, researchers, parents, teachers, and school administrators. One has to wonder about the motivation behind language policies that clearly affect Latino ELLs' opportunity to learn mathematics. Moll and Ruiz (2002) call for "educational sovereignty" as a way to challenge "the arbitrary authority of the white power structure to determine the essence of education for Latino students" (p. 368). Part of this educational sovereignty includes valuing Latino cultural, social, and linguistic backgrounds. CEMELA's research in classrooms and after-school settings calls for policymakers to address the impact of social bias toward bilingualism and toward Spanish as an important step toward improving Latinos' mathematics education. We should make efforts to recognize the negative cognitive, social, and psychological effects of subtractive-schooling approaches that restrict using students' first language and damp the positive cognitive effects of home-language maintenance.

What will it take for schools to build on the mathematical knowledge that ELLs have and to recognize the resources that ELLs bring by being bilingual and bicultural? Researchers and teacher educators need to do more in teachers' preparation and professional development programs, in order to engage in-service and preservice teachers in reflecting on the role of language and culture in mathematics teaching and learning with Latino and Latina students. Our work shows that engaging teachers in these conversations is not an easy task: teachers often turn to the culture of poverty to explain students' low achievement (Civil 2009). We need to provide more opportunities for teachers to appreciate and value the resources that bilingual students draw on when doing mathematics, that is, their use of their first language, gestures, and code switching (Zahner and Moschkovich, in press), as well as examples of ELLs' successful participation in mathematical discussions.

We also need to provide more opportunities for parents and teachers to interact in children's mathematics education in formats other than the parent-teacher conference. Many misperceptions occur between parents and teachers about their respective roles in the teaching and learning process. CEMELA's outreach to parents, in which we bring together preservice teachers, teachers, parents, children, and university researchers, allows for parents and teachers to learn from one another. Through these outreach efforts, teachers and school administrators can learn from the parents, not only about parents' expectations for their children's schooling but also parents' approaches to school mathematics and their uses of mathematics in everyday life. These opportunities for dialogue may improve communication around, for example, homework, and they may make parents and teachers more aware of issues such as placement. In these CEMELA opportunities, bilingualism becomes a tool to mediate learning, allowing parents to share the academic arena with their children. Although in Arizona English is the prevalent school language and Spanish may be the one in the home, workshops for parents and children show a back-and-forth between both languages, allowing a better understanding of reasons for language choice (Menéndez, Civil, and Mariño 2009). These activities with parents can offer in-service and preservice teachers ideas on how to build on their students' linguistic and cultural backgrounds, as well as how to challenge or expand their prior thinking about Latino parents and students. To quote a preservice teacher,

> More than anything, I have learned about the parents. I never thought they would be so involved and dedicated to the education of their kids. Working with CEMELA has also made me realize the importance of culture and language and how key it is to the success of students. If they feel disconnected at all or out of place it obviously affects their education.

REFERENCES

Acosta-Iriqui, Jesús, Marta Civil, Javier Díez-Palomar, Mary Marshall, and Beatriz Quintos-Alonso. "Conversations around Mathematics Education with Latino Parents in Two Borderland Communities: The Influence of Two Contrasting Language Policies." In *Latinos and Mathematics Education: Research on Learning and Teaching in Classrooms and Communities*, edited by Kip Téllez, Judit Moschkovich, and Marta Civil. Greenwich, Conn.: Information Age Publishing, in press.

Anhalt, Cynthia O., Matthew Ondrus, and Virginia Horak. "Insights from Middle School Teachers' Participation in a Mathematics Lesson in Chinese." *Mathematics Teaching in the Middle School* 13 (August 2007): 18–23.

Callahan, Rebecca, Lindsey Wilkinson, and Chandra Muller. "School Context and the Effect of ESL Placement on Mexican-Origin Adolescents' Achievement." *Social Science Quarterly* 89 (January 2008): 177–98.

Civil, Marta. "Building on Community Knowledge: An Avenue to Equity in Mathematics Education." In *Improving Access to Mathematics: Diversity and Equity in the Classroom*, edited by Na'ilah Suad Nasir and Paul Cobb, pp. 106–17. Mahwah, N.J.: Lawrence Erlbaum Associates, 2007.

———. "Language and Mathematics: Immigrant Parents' Participation in School." In *Proceedings of the Joint Meeting of PME 32 and PME-30*, edited by Olimpia Figueras, José Luis Cortina, Silvia Alatorre, Teresa Rojano, and Armando Sepúlveda. Vol. 2, pp. 329–36. Morelia, Michoacan, Mexico: Centro de Investigación y de Estudios Avazados del Instituto Politéchnico Nacional—Universidad Michoacana de San Nicolás de Hidalgo, 2008a.

———. "Mathematics Education, Language, and Culture: Ponderings from a Different Geographic Context." In *Crossing Divides: Proceedings of the 32nd Annual Conference of the Mathematics Education Research Group of Australasia*, edited by Roberta Hunter, Brenda Bicknell, and Tim Burgess, pp. 129–36. Palmerston North, New Zealand: Massey University, 2009.

———. "Mathematics Teaching and Learning of Immigrant Students: A Look at the Key Themes from Recent Research." Paper presented at the 11th International Congress of Mathematics Education Survey Team 5: Mathematics Education in Multicultural and Multilingual Environments, Monterrey, Mexico, July 2008b.

———. "When the Home Language Is Different from the School Language: Implications for Equity in Mathematics Education." Paper presented at the Topic Study Group 31, Language and Communication in Mathematics Education, at the 11th International Congress of Mathematics Education, Monterrey, Mexico, July 2008c.

Civil, Marta, and Rosi Andrade. "Collaborative Practice with Parents: The Role of the Researcher as Mediator." In *Collaboration in Teacher Education: Examples From the Context of Mathematics Education*, edited by Andrea Peter-Koop, Vania Santos-Wagner, Chris Breen, and Andy Begg, pp. 153–68. Boston: Kluwer Academic Publishers, 2003.

Civil, Marta, and José María Menéndez. "Impressions of Mexican Immigrant Families on Their Early Experiences with School Mathematics in Arizona." In *Transnational and Borderland Studies in Mathematics Education*, edited by Richard S. Kitchen and Marta Civil, pp. 47–68. New York: Routledge, 2011.

Civil, Marta, and Núria Planas. "Latino/a Immigrant Parents' Voices in Mathematics Education." In *Immigration, Diversity, and Education*, edited by Elena L. Grigorenko and Ruby Takanishi, pp. 130–50. New York: Routledge, 2010.

Combs, Mary Carol, Carol Evans, Todd Fletcher, Elena Parra, and Alicia Jiménez. "Bilingualism for the Children: Implementing a Dual-Language Program in an English-Only State." *Educational Policy* 19 (November 2005): 701–28.

Cummins, Jim. *Language, Power, and Pedagogy: Bilingual Children in the Crossfire*. Tonawanda, N.Y.: Multilingual Matters, 2000.

González, Norma, Luis C. Moll, and Cathy Amanti, eds. *Funds of Knowledge: Theorizing Practice in Households, Communities, and Classrooms*. Mahwah, N.J.: Lawrence Erlbaum Associates, 2005.

Hernandez, Donald J., Nancy A. Denton, and Suzanne E. Macartney. "Children of Immigrants and the Future of America." In *Immigration, Diversity, and Education*, edited by Elena L. Grigorenko and Ruby Takanishi, pp. 7–25. New York: Routledge, 2010.

Khisty, Lena Licón, and Kathryn B. Chval. "Pedagogic Discourse and Equity in Mathematics: When Teachers' Talk Matters." *Mathematics Education Research Journal* 14 (December 2002): 154–68.

Khisty, Lena Licón, and Craig Willey. "The Politics of Language and Schooling in the Mathematics Education of Bilingual Chicana/o Students." Paper presented at Topic Study Group 33, Mathematics Education in a Multilingual and Multicultural Environment, at the 11th International Congress of Mathematics Education, Monterrey, Mexico, July 2008.

Lappan, Glenda, James T. Fey, William M. Fitzgerald, Susan N. Friel, and Elizabeth Difanis Phillips. *Filling and Wrapping: Three-Dimensional Measurement-Connected Mathematics 2*. Boston: Pearson/Prentice Hall, 2006.

Menéndez, José María, Marta Civil, and Verónica Mariño. "Latino Parents as Teachers of Mathematics: Examples of Interactions Outside the Classroom." Paper presented at the annual meeting of the American Educational Research Association, San Diego, Calif., April 2009.

Moll, Luis C., and Richard Ruiz. "The Schooling of Latino Children." In *Latinos: Remaking America*, edited by Marcelo M. Suárez-Orozco and Mariela M. Páez, pp. 362–74. Berkeley, Calif.: University of California Press, 2002.

Moschkovich, Judit. "A Situated and Sociocultural Perspective on Bilingual Mathematics Learners." *Mathematical Thinking and Learning* 4, nos. 2–3 (2002): 189–212.

———. "Using Two Languages While Learning Mathematics." *Educational Studies in Mathematics* 64 (February 2007): 121–44.

National Council of Teachers of Mathematics (NCTM). "Equity in Mathematics Education: A Position Statement of the National Council of Teachers of Mathematics." Reston, Va.: NCTM, 2008.

Planas, Núria. and Marta Civil. "El Aprendizaje Matemático de Alumnos Bilingües en Barcelona y Tucson." Manuscript submitted for publication, 2009.

Tate, William F. *Access and Opportunity to Learn Are Not Accidents: Engineering Mathematical Progress in Your School.* Greensboro, N.C.: Southeast Eisenhower Regional Consortium for Mathematics and Science at SERVE, 2005.

Turner, Erin E., Sylvia Celedón-Pattichis, and Mary Marshall. "Cultural and Linguistic Resources to Promote Problem Solving and Mathematical Discourse among Hispanic Kindergarten Students." In *Promoting High Anticipation and Success in Mathematics by Hispanic Students: Examining Opportunities and Probing Promising Practices,* TODOS: Mathematics for ALL Monograph # 1, edited by Richard S. Kitchen and Edward Silver, pp. 19–42. Washington, D.C.: National Education Association Press, 2008.

Turner, Erin E., Maura Varley Gutiérrez, Ksenija Simic-Muller, and Javier Díez-Palomar. " 'Everything Is Math in the Whole World': Integrating Critical and Community Knowledge in Authentic Mathematical Investigations with Elementary Latina/o Students." *Mathematical Thinking and Learning* 11 (July 2009): 136–57.

Valdés, Guadalupe. *Learning and Not Learning English: Latino Students in American Schools.* New York: Teachers College Press, 2001.

Valenzuela, Angela. *Subtractive Schooling: U.S.-Mexican Youth and the Politics of Caring.* Albany, N.Y.: State University of New York Press, 1999.

Varley Gutiérrez, Maura C. " 'I Thought This U.S. Place Was Supposed to Be about Freedom': Young Latinas Speak to Equity in Mathematics Education and Society." Ph.D. diss., University of Arizona, 2009.

Wright, Wayne E. "The Political Spectacle of Arizona's Proposition 203." *Educational Policy* 19, no. 5 (2005): 662–700.

Zahner, William, and Judit Moschkovich. "Bilingual Students Using Two Languages during Peer Mathematics Discussions: ¿Qué Significa? Estudiantes Bilingües Usando Dos Idiomas en sus Discusiones Matemáticas: What Does it Mean?" In *Latinos and Mathematics Education: Research on Learning and Teaching in Classrooms and Communities,* edited by Kip Téllez, Judit Moschkovich, and Marta Civil. Greenwich, Conn.: Information Age Publishing, in press.

Elementary Mathematics Specialists: A Merger of Policy, Practice, and Research

Patricia F. Campbell

URRENTLY, a crucial domain at the intersection of research, practice, and policy in mathematics education is students' achievement as measured on high-stakes mathematics tests. The mathematical content assessed on these measures is only a subset of that specified in a state's or school district's curriculum framework. And, ideally, the mathematics content specified in a framework is an essential component of the legitimate trajectory of rich mathematics that knowledgeable teachers target in their classrooms, using appropriate instructional strategies and resources to foster students' learning. But in public schools today, from a political and policy perspective, the coin of the realm is advancing students' achievement as measured on states' high-stakes mathematics tests, not the richer domain of mathematics that knowledgeable teachers may wish to explore with their students. Admittedly, this is limiting (Valli et al. 2008). Nevertheless, public schools remain a rich setting for mathematics education research addressing teaching and learning. Further, mathematics education research has a responsibility to design and carry out work that may guide mathematics education practice in the current context of high-stakes testing.

Frequently, the role of research is investigating hypotheses, such as whether a seemingly good idea or treatment works or under what conditions it does or does not work. As such, the intent of research is to contribute to knowledge. Schön (1987) noted that "[i]n the varied topography of professional practice, there is a high, hard ground overlooking a swamp. On the high ground, manageable problems lead themselves to solution through the application of research-based theory and technique. In the swampy lowland, messy, confusing problems defy technical solution" (p. 3). Mathematics education research investigating the effectiveness of school- or district-level efforts to advance students' achievement in this era of shifting policy demands resides in the swamp. Nevertheless, to the extent possible, education research requires a "clean"

The project described in this chapter was developed with the support of National Science Foundation grant no. ESI-0353360. The statements and findings herein reflect the opinions of the author and not necessarily those of the National Science Foundation.

implementation of a treatment, documenting the variables at play in a school or classroom and how these factors influenced the conduct of the treatment or idea being studied.

When education is a successful enterprise, it is marked by collaboration between teachers, principals, students, parents, and district-level administrators, as well as attention to district and state policy specifying assessment, curriculum, and licensure. Research investigating the effectiveness of ideas or treatments defining change in some aspect of mathematics education in an effort to advance students' achievement requires collaboration among, and attention to, most of these constituents. Attention to policy is crucial, not only because it may determine or manipulate implementation, but also because, although the entire research endeavor may prove that a good idea works, without a subsequent change in policy, the idea remains just an idea. Without supportive policy, proven good ideas are not sustained and applied in educational settings in the future.

To illustrate how these constituents may come together over time to influence policy and to improve mathematics education, consider a collaborative endeavor addressing elementary mathematics specialists.

Elementary Mathematics Specialists in Virginia

The Virginia Mathematics Coalition was organized in 1990 as a nonprofit organization composed of university and college mathematicians and mathematics educators who joined with leaders from school districts, business, industry, and public policy centers from across Virginia, all with the goal of invigorating grades pre-K–16 mathematics education in Virginia. In 1991, the Coalition and the Virginia Department of Education received funding from the National Science Foundation (NSF) for a state systemic initiative; one of the components of that reform effort was a professional development series designed for lead teachers of mathematics in elementary schools. After the systemic funding ended, the potential of the lead-teacher idea continued to intrigue the collaborators, although they were cognizant of the need to address the conflicting demands reported by their lead teachers. How could a single person be a classroom teacher responsible for students' instruction and still be available and prepared to serve as a school-based resource for other teachers? What if these teacher leaders had a stronger preparation and background in mathematics content, instructional strategies, and school leadership and were assigned full time as elementary mathematics specialists?

By 1998, the Coalition had expanded into the Virginia Mathematics and Science Coalition, bringing scientists, science educators, and more corporate support into the collaboration. In time, one Coalition member was elected to the General Assembly of Virginia. Yet, this expanded entity remained committed to the idea of elementary mathematics specialists. Indeed, a number of Coalition members were quite explicit about one of their goals for the Coalition—place well-prepared, well-supported mathematics specialists in every elementary school in Virginia. But, more and more often, when Coalition members discussed positioning elementary mathematics specialists in schools, educational policy makers and school district leaders asked whether elementary mathematics specialists affected students' performance on the Standards of Learning (SOL), Virginia's high-stakes achievement tests administered as required under No Child Left

Behind legislation. Without an answer to this question, school district administrators were reluctant to proceed with the policy modifications that developing and financing these positions would require.

To address this persistent question and programmatic policy questions that would arise in the process of defining these positions, the Coalition joined with leaders from the Virginia Council of Teachers of Mathematics, the Virginia Council for Mathematics Supervision, the mathematics staff of the Virginia Department of Education, and a mathematics education researcher at the University of Maryland. The resulting partnership sought grant funding for an elementary mathematics specialist preparation program and for a research study investigating the impact of graduates of that program on students' achievement. The enterprise added new partners and defined new intermediate goals to answer questions being raised by school district leaders and policymakers.

The Elementary Mathematics Specialist Project

In 2004, the NSF funded a collaborative project involving four universities and five school districts to accomplish the following:

- Develop integrated mathematics content courses that would increase the participants' mathematical content knowledge, deepen their understanding of school mathematics concepts and skills, enhance their specialized content knowledge for teaching mathematics (Ball, Thames, and Phelps 2008), and augment the participants' pedagogical content knowledge for mathematics

- Develop leadership-coaching courses that would address the participants' understanding of current research on teaching and learning mathematics, develop participants' leadership and facilitation skills, and support their emergent expertise as change agents and coaches

- Conduct a randomized control trial investigating the impact of participants' assignment as an elementary mathematics specialist on students' achievement and teachers' beliefs

- Identify and analyze policy, legislative, and regulatory issues regarding implementing the mathematics specialist program at state and local levels, and inform and advise the grant management team regarding policy decisions in these areas

Thus, although the long-term goal of positioning and supporting well-prepared elementary mathematics specialists across the state persisted, the focus shifted to establishing mechanisms for developing specialists, to researching their impact, and to identifying and understanding policy issues that would potentially affect the feasibility of achieving that long-term goal.

Management and Responsibility

The project team distributed responsibility for the project's primary components among its members according to expertise. Mathematicians, mathematics educators, and school-district supervisors or coordinators for mathematics addressed course development and delivery; a mathematics education researcher carried out the research protocol; and representatives from the Commonwealth Educational Policy Institute monitored public, state-level policy discussions. A management team was established, made up of at least one mathematician or mathematics educator from each of the universities, the mathematics education researcher, mathematics supervisors from two of the school districts, the state mathematics supervisor, and two of the policy team members. This team met bimonthly over the course of the grant to review progress, address emergent issues, and determine midcourse corrections. One of the mathematicians served as the principal investigator, maintaining communication between those focusing on course development and delivery, research, and policy. A school district mathematics supervisor or coordinator and either a university mathematics educator or a university mathematician team-taught each specialist preparation course, further establishing lines of trust and routes for information exchange over the course of the collaboration.

Policy Levers

Three domains emerged as crucial settings for policy decisions affecting elementary mathematics specialists in Virginia. These were (1) clarifying the role of an elementary mathematics specialist, (2) defining coursework that would prepare would-be specialists for their new role, and (3) communicating and interacting with policymakers, which require an understanding of how educational policy is proposed, reviewed, and prescribed in Virginia.

The first policy-driven challenge the project faced was defining the elementary mathematics specialist position. What were an elementary mathematics specialist's responsibilities? The immediate need for a "job description" arose, in order to communicate the expectations for the position to partnering school districts and to begin mapping the terrain for preparation courses. This definition proved essential, because any future licensure for elementary mathematics specialists would require specifying their role and their needed qualifications.

Clarifying the role of the elementary mathematics specialist led to the second domain—defining the content of a series of preparation courses that would fit into the framework of approved master's degree programs already offered in Virginia's differing higher-education institutions. The project did not require participants to acquire a master's degree, partly because a number of the initial applicants already had earned one. However, the participants were required to complete the courses that the project developed and offered. Because the project-developed mathematics content and leadership-coaching courses satisfied a number of categories of requirements in approved master's degree frameworks, participants who did not have a master's degree could thus use those courses in their plans of study, along with additional coursework that their degree-granting institutions prescribed, to earn a degree.

This second domain was also important because, if Virginia ever moved to licensing elementary mathematics specialists in the future, the Virginia State Department of Education would need programmatic requirements to determine whether an applicant qualified for certification. By designing a coherent, encompassing series of mathematics content and leadership-coaching courses, the project hoped to create a model that the Virginia State Department of Education might refer to when stipulating the coursework or proficiency required for elementary mathematics certification.

The project's policy team assumed the somewhat nebulous but vitally important task of the third domain—establishing policy communications and connections. The project management team's mathematicians and educators needed to understand Virginia's education governance and policy-making structures, as well as Virginia's legislative and regulatory processes, and to know what Virginia Board of Education actions would influence the climate for support of, or opposition to, creating the elementary mathematics specialist position. Further, the project management team needed to (1) be alert to opportunities for providing evidence of the need for elementary mathematics specialists to state policymakers and (2) learn how to do so in ways that these policymakers consider credible. For example, the policy team met with staff members of a Virginia state senator, who also was a member of the Mathematics and Science Coalition, to supply talking points and letters of support for elementary mathematics specialist positions. This contributed to a resolution, eventually passed by the General Assembly of Virginia, requesting that the Virginia Board of Education address the inclusion of an endorsement for mathematics specialists in its pending revisions of *Licensure Regulations for School Personnel.* The policy team also helped teachers and mathematicians refine testimony for presentation at Board of Education hearings, as well as meetings with Virginia State Department of Education officials where teachers and mathematicians advocated for the licensure of mathematics specialists. Mathematicians on the project management team made presentations to the Academic Advisory Committee and the Board of Directors of the State Council for Higher Education for Virginia regarding the preparation courses, to promote their inclusion in a flexible, statewide master's degree program.

Policy communication and connections were not limited to interactions at state level. Indeed, policy team members interviewed school board members, district-level administrators (e.g., superintendents, directors of instruction, supervisors of mathematics), and principals from the school districts participating in the research study to identify pertinent local policy issues regarding the implementation of elementary mathematics specialists. Further, members gave a presentation regarding the role of elementary mathematics specialists at the annual meeting of the Virginia School Board Association. They also published information articles in the newsletters of the Virginia Association of Elementary School Principals and the Virginia Association of School Superintendents.

Foundations

If a school's mathematics program is going to support students' learning, mathematics specialists and elementary school teachers must acquire and effectively use the knowledge of mathematics and pedagogy. The intention of the five mathematics content courses was to address the specialists' mathematics content and pedagogical knowledge using resources that would challenge the specialists to understand mathematics and not just to apply procedures by rote (e.g., Carpenter, Franke, and Levi [2003]; Fosnot and Dolk [2002]; Lamon [1999]; Schifter, Bastable, and Russell [1999a], [1999b]). The first leadership-coaching course focused on knowledge and perspectives for teaching mathematics, pushing the participants to reflect on their own teaching and to understand the trajectory of mathematics curriculum expectations across grades K–5 (e.g., Kilpatrick, Swafford, and Findell [2001]; National Council of Teachers of Mathematics [2000]). The second leadership-coaching course directly addressed coaching and analyzing instruction, with particular focus on the connections between instructional moves and evidence of students' learning (e.g., Ma [1999]; West and Staub [2003]). Participants attended a summer workshop prior to their first year of placement in a school as a specialist. This workshop considered some of the challenges that whole-school specialists experience, such as dealing with principals, balancing multiple responsibilities, and setting priorities within time constraints. Some districts also offered periodic professional development sessions for specialists over time.

Evaluation Design and Outcomes

Formative evaluation played a vital role in the elementary mathematics specialist project. Horizon Research, Inc., observed debriefing and planning sessions for varying offerings of the courses, observed selected sessions of the courses, interviewed approximately half the course participants each year, and administered a post-course survey to participants for each course. The aggregated findings from these surveys, as well as selected anonymous comments from participants, provided the substance for an annual report that the Horizon Research evaluators submitted to the project management team each year. This report described useful information gleaned during the review and modification of the mathematics content and the leadership-coaching courses. It also gave the management team some insights regarding the specialists' experiences after being placed in a school. These insights were useful when revising the leadership-coaching courses and provided discussion issues for annual, overnight, project-organized, professional development retreats for the specialists and district-level mathematics supervisors.

Horizon Research's summative evaluation report (Wickwire, Smith, and Moffett 2009) emphasized three outcomes related to specialists' preparation. First, the mathematics courses not only enhanced participants' content knowledge and sense of preparedness, but also influenced participants' perceptions of mathematics teaching. In particular, participants noted that they "wanted their students to develop a more conceptual understanding of content and understand 'the why' behind different procedures or algorithms," and they wanted their classrooms to be locations where students could "discuss their thinking and help each other to build more connections and meaning" (p. 4). Second, the leadership-coaching courses strengthened specialists' perceptions of themselves as leaders and coaches. Finally, the specialists noted that they valued the opportunities the courses and project offered for interactions with one another, because these interactions contributed to their learning.

In addition to findings addressing the specialist preparation courses, the evaluation reports from Horizon Research, Inc., offered two extremely important insights into specialists' transition from the role of classroom teaching to that of assuming coaching and programmatic responsibilities in a school. First, principals' priorities and support varied, and principals strongly influenced specialists' role and credibility. If a supportive environment existed, then a specialist and his or her school's instructional and administrative staffs learned and worked together. Second, the expectations set for specialists varied widely across schools. For example, although the need to raise students' SOL scores was a fact for every specialist and school, some specialists essentially assumed part-time, assessment-coordinator roles as they collected, analyzed, and reported test data across subjects and grades. Similarly, when a specialist replaced a Title I mathematics resource teacher in a school, both teachers and the principal frequently expected the specialist would assume the responsibility of providing regularly scheduled, targeted instruction outside the classroom to identified students. Resetting those expectations was a challenge.

Research Design and Outcomes

For the project's randomized treatment-control trial, each of the five Virginia school districts collaborating in the project—representing urban, urban-edge, and rural-fringe communities—identified one or more triples of schools with comparable student demographics and comparable traditions of students' performance on state mathematics assessments. They chose a total of 12 triples of schools, rather than pairs, in order to yield comparable school placement sites for two differing cohorts of mathematics specialists while maintaining corresponding control sites. The project researcher randomly chose one school from each of the 12 triples. A cohort of 12 specialists completed five mathematics content courses and one leadership-coaching course prior to placement in an identified treatment school. These specialists completed the second leadership-coaching course during their first year of service in these schools. A second cohort of 12 specialists enrolled in a revised offering of the same content and leadership courses. After they completed those courses, their districts placed them in one of the two remaining schools in each of the original triples. The project researcher randomly selected these Cohort 2 treatment sites.

In order to determine whether these elementary mathematics specialists affected students' mathematics achievement in Virginia, the project completed two analyses of students' SOL scale scores drawn from grades 3, 4, and 5 at the 36 treatment and control schools over three years. The treatment-versus-control analysis compared three years of mathematics achievement scores of students in the control schools to three years of scores of students in treatment schools with a specialist. This analysis on data collected in the study's third year identified whether the achievement scores of students in the treatment schools were from the three years of data from Cohort 1 schools or from the one year of data from the Cohort 2 schools. A second, cohort-by-year-versus-control analysis compared three years of scores of students in the control schools to those of students in the treatment schools. This second analysis noted whether the achievement scores of treatment-school students' came from (1) the first (Cohort 1 Year 1), second (Cohort 1 Year 2), or third (Cohort 1 Year 3) year of specialist placement in a Cohort 1 school; or (2) the first year of specialist placement in a Cohort 2 school (Cohort 2 Year 3) during the study's third year.

The treatment-versus-control analysis indicated a statistically significant difference in the SOL mathematics achievement scores of students in the schools with the Cohort 1 elementary mathematics specialists as compared to students in the control schools. This analysis of three years of data from Grade 3, from Grade 4, and from Grade 5 revealed that at each of these grades, the students in the schools with the specialists had significantly higher scores. This analysis analyzed the three years of data from each of these three grades separately.

This second treatment-versus-control analysis also indicated that the Cohort 2 Year 3 variable, representing a first-year specialist's placement during the third year of the study, was not significant at any of the tested grades. During the first year of placement, Cohort 2 specialists did not significantly impact the SOL mathematics scores of students in their schools, as compared to those of students in Cohort 1 or control schools.

The cohort-by-year-versus-control analysis permitted an examination of whether this difference in findings between cohorts reflected (1) the differing amounts of time the elementary mathematics specialists in the two cohorts had to work with teachers and the school mathematics program or (2) a cohort effect. Because this analysis separated the Cohort 1 data into three sets, one for each year of specialist placement, increased standard errors associated with this more conservative analysis result. Although this influenced the likelihood of statistical significance, the purpose for conducting this second analysis's purpose was to identify any patterns of achievement, not to identify significant differences.

The cohort-by-year-versus-control analysis indicated that, although the SOL mathematics scores of students in the Cohort 1 schools were, on average, numerically higher than those of students in the control schools during the Cohort 1 specialists' first year of placement, this difference was not statistically significant at any grade. This pattern of higher SOL mathematics achievement scores in Cohort 1 schools, compared to those from the control schools, persisted into the second year of placement, with a significant difference noted in grades 4 and 5 but not in grade 3. During the third year of placement, this pattern of greater scores in Cohort 1 schools across all three grades continued, with a statistically significant difference noted only in grade 5. As expected, increased standard errors in grades 3 and 4 influenced significance in the third-year data.

Interpreted together, these two analyses found no significant difference in students' achievement on Virginia's high-stakes mathematics assessment between schools with and without specialists in the first year of specialist placement in either cohort. The pattern of achievement was as follows:

- An increase in SOL scores in the schools with specialists, as compared to schools without specialists in Year 1;

- A greater increase in the difference in SOL scores in Year 2 between schools with and without specialists, favoring the schools with specialists; and

- A continuing increase in the difference in SOL scores in Year 3, favoring schools with a specialist in her third placement year.

The increase in achievement in Years 2 and 3 in the Cohort 1 schools drove the overall statistically significant difference in students' achievement, favoring Cohort 1 schools over the control schools and over Cohort 2 schools in the three-year, treatment-versus-control analysis.

Policy Outcomes

This effort led to two crucial policy outcomes, one involving the specialist preparation courses and the other addressing licensure.

The State Council for Higher Education for Virginia approved the Virginia Mathematics and Science Coalition Statewide Masters Programs, though which participating Virginia universities may collaborate to offer coursework leading to master's degrees in combinations of science, mathematics, education, and leadership for science and mathematics teachers. Although individual universities offer these graduate degrees and courses, and students need to be admitted to a participating university, the participating universities develop the courses collaboratively, and each participating university will automatically accept them as nontransfer credits. The first of the approved statewide master's programs, as submitted by six Virginia universities, incorporated the project's five mathematics content courses and two leadership-coaching courses.

Teachers interested in the preparatory coursework for a mathematics specialist position may thus take advantage of course offerings at colleges and universities throughout Virginia, instead of having to wait for a course offered at the institution where he or she is pursuing a master's degree. Also, higher-education institutions can rotate course offerings, permitting sufficient registration to accrue without restricting any candidate's progress toward degree completion.

The second policy outcome was the approval of a grades K–8 mathematics specialist licensure endorsement by the state of Virginia in September 2007. A brief recounting of the process by which the endorsement's authorization proceeded illustrates the importance of monitoring the activity of state-level legislative and regulatory activity.

In February 2005, the General Assembly of Virginia requested that the Virginia Board of Education include an endorsement for mathematics specialists in its consideration of revisions for licensure. Subsequently, the Virginia Board of Education requested that Virginia's Advisory Board on Teacher Education and Licensures (ABTEL), an advisory group available to the Board of Education, develop and examine criteria for licensure and make recommendations. ABTEL commissioned the Commonwealth Educational Policy Institute to prepare a detailed report describing the status of mathematics specialists in Virginia and the associated policy and regulatory issues. The institute referred to the project's work for a sample position description and a sample program designed to prepare specialists for needed competencies. In June 2005, the Board of Education considered the changes to regulations for licensure that ABTEL had submitted, and the board authorized the Virginia Department of Education to proceed with the specification of an endorsement and programmatic requirements for grades K–8 mathematics. Although the project focused solely on grades K–5, the final recommendation included the middle grades.

Opportunity for public comment after the posting of proposed endorsement regulations and programmatic requirements proceeded in fall 2006. The Virginia Mathematics and Science Coalition worked with the project policy team to secure advocates for the endorsement who either spoke at public hearings or submitted written statements of support. Simultaneously, the Virginia Board of Education held public hearings addressing changes in the Standards of Quality (SOQ), the public code that specified requirements for local school boards. The SOQ's purview includes, among other components, instructional programs and support personnel. The existing SOQ specified support for reading, but not for mathematics. The Coalition publicized the need for speakers for these hearings, who then advocated for mathematics in the revised SOQ and promoted mathematics specialists as a means of incorporating enhanced instructional strategies for mathematics in school programs. The Board of Education recommended approval for a grades K–8 mathematics specialist endorsement in 2007 as well as the revised SOQ, and the Virginia legislature enacted the revised SOQ. In September 2007, the governor signed the authorization (1) establishing licensure for elementary mathematics specialists through endorsement and (2) modifying the SOQ.

Lessons Learned from the Elementary Mathematics Specialist Project

This project brought together participants with expertise from a variety of venues that, by forging connections that accessed their distinct talents, influenced state-level policy determining mathematics education and generated research guiding mathematics education practice. As such, this effort offers a few principles that may serve as guides for future collaborations seeking to address policy.

Be Willing to Learn

This project illustrates the importance of partners sharing a common goal, as well as the advantage of bringing varying areas of expertise to the table while respecting the variety of knowledge that different team members contribute. The mathematics educators and mathematicians in this enterprise had worked together previously and were well known in Virginia. Yet they knew that further efforts to influence decision makers at state and local levels would be stymied unless those efforts addressed policy and research investigating how the project's aim would affect students' achievement. But, in the words of one of the mathematicians, they "didn't have a clue" about how to "do" control-trial research or policy work. As a result, the project expanded its personnel to include this chapter's author and a policy team. At the same time, everyone knew that there were no guarantees that the research would indicate a positive effect on students' achievement or that licensure compatible to their vision would materialize.

Embed the Enterprise

A second characteristic of this endeavor was that it was embedded with the teachers, specialists, mathematics supervisors, and principals across the partnered school districts, as well as the Virginia Mathematics and Science Coalition. This allowed multiple constituencies the issue of elementary mathematics specialists repeatedly to many audiences, and to garner support for legislative and regulatory actions. In addition, by keeping lines of communication open and being transparent about research requirements and the regulatory process, the project could proceed while audiences waited for the results.

Collaboration, Not Competition

Policymakers, researchers, professional developers, mathematicians, mathematics educators, administrators, and practitioners have differing perspectives. Throughout this endeavor, the potential implications of a decision's impact on research, policy, and practice were frequently contradictory and sometimes elusive. Thus, everyone engaged in the project needed to share a noncompetitive sense of mutual obligation and responsibility. This hallmark of our elementary mathematics specialist enterprise was why the effort eventually paid dividends.

REFERENCES

Ball, Deborah Loewenberg, Mark Hoover Thames, and Geoffrey Phelps. "Content Knowledge for Teaching: What Makes It Special?" *Journal of Teacher Education* 59 (November/December 2008): 389–407.

Carpenter, Thomas P., Megan Loef Franke, and Linda Levi. *Thinking Mathematically: Integrating Arithmetic and Algebra in Elementary School*. Portsmouth, N.H.: Heinemann, 2003.

Fosnot, Catherine Twomey, and Maarten Dolk. *Young Mathematicians at Work: Constructing Fractions, Decimals, and Percents*. Portsmouth, N.H.: Heinemann, 2002.

Kilpatrick, Jeremy, Jane Swafford, and Bradford Findell, eds. *Adding It Up: Helping Children Learn Mathematics*. Washington, D.C.: National Academies Press, 2001.

Lamon, Susan J. *Teaching Fractions and Ratios for Understanding: Essential Content Knowledge and Instructional Strategies for Teachers*. Mahwah, N.J.: Lawrence Erlbaum Associates, 1999.

Ma, Liping. *Knowing and Teaching Elementary Mathematics: Teachers' Understanding of Fundamental Mathematics in China and the United States*. Mahwah, N.J.: Lawrence Erlbaum Associates, 1999.

National Council of Teachers of Mathematics (NCTM). *Principles and Standards for School Mathematics*. Reston, Va.: NCTM, 2000.

Schifter, Deborah, Virginia Bastable, and Susan Jo Russell. *Making Meaning for Operations: Casebook*. Parsippany, N.J.: Dale Seymour Publications, 1999a.

———. *Making Meaning for Operations: Facilitator's Guide*. Parsippany, N.J.: Dale Seymour Publications, 1999b.

Schön, Donald A. *Educating the Reflective Practitioner*. San Francisco: Jossey-Bass, 1987.

Valli, Linda, Robert G. Croninger, Marilyn J. Chambliss, Anna O. Graeber, and Daria Buese. *Test Driven: High-Stakes Accountability in Elementary Schools*. New York: Teachers College Press, 2008.

West, Lucy, and Fritz C. Staub. *Content-Focused Coaching: Transforming Mathematics Lessons*. Portsmouth, N.H.: Heinemann, 2003.

Wickwire, Murray E., P. Sean Smith, and Gwen E. Moffett. *NSF Institute: Preparing Virginia's Math Specialists Final Report*. Chapel Hill, N.C.: Horizon Research, Inc., 2009.

Transforming East Alabama Mathematics (TEAM-Math): Promoting Systemic Change in Schools and Universities

W. Gary Martin
Marilyn E. Strutchens
Stephen Stuckwisch
Mohammed Qazi

A LABAMA has a long history of low achievement in grades K–12 mathematics. The National Assessment of Educational Progress (NAEP) (National Center for Educational Statistics [NCES] 2009), also known as "The Nation's Report Card," has shown Alabama at or near the bottom among the states on the national mathematics assessment at both grades 4 and 8 since state-level comparisons began in 1992. Moreover, significant gaps in mathematics achievement have been identified among various demographic groups in Alabama on both NAEP and mandated state assessments, for all grade levels going back as far as such data was reported (NCES 2009; Alabama State Department of Education [ALSDE] 2009b). These gaps include differences between white and African American students, the two major groups in Alabama by race and ethnicity; between students who do and do not participate in the National School Lunch Program; and between students who are and are not identified for special education.

Beyond the statewide issues in mathematics achievement, the eastern region of Alabama has additional needs and challenges. Analyses of state-level data show both an overall lower level of mathematics achievement and more pronounced gaps in achievement among demographic groups in the school districts of east Alabama than in the state as a whole. Moreover, the small size of the districts in the region, their general lack of resources, and the dispersed population add barriers to improving mathematics education: the region, spread over a 100-mile radius, consists of a series of small school districts with an average size of 4000 students each.

In 2002 a group of faculty members from Auburn University's Department of Mathematics and Statistics met with two relatively new mathematics education faculty members in the

Department of Curriculum and Teaching. They all recognized that they had common concerns about addressing the problems in mathematics education in east Alabama. As a result, the College of Science and Mathematics provided funding to establish a planning team to explore ways to respond to this challenge. The group invited a number of speakers from innovative programs aimed at enhancing grades K–12 mathematics programs to Auburn University's campus, to present their research and development programs and meet with the planning group. The speakers included Patricia F. Campbell (2010), principal investigator for Mathematics Applications and Reasoning Skills, a systemic change project with Baltimore City Public Elementary Schools; and William Bush from the Appalachian Collaborative Center for Learning. As a result of the information exchange, the planning team decided that a systemic change model based on substantive university-school partnerships would best meet the region's needs. They took steps through the end of 2002 to engage with potential partners. Members of the planning team already collaborated with Tuskegee University on other endeavors, so faculty from that institution joined the partnership. Efforts to engage other higher education partners in the area were ultimately unsuccessful.

A number of teacher leaders and administrators from school districts in the region attended the talks held by the planning team and supported the proposed directions. A local superintendent, who had previously collaborated with the mathematics faculty on a distance-learning project, joined the planning team. He helped to set up a meeting with the superintendents of the eleven school districts within a 50-mile radius of his own. All eleven signed on to the partnership at the initial meeting. In early 2003, an additional school district just beyond the 50-mile radius asked to join the partnership. In 2005, the partnership invited three additional districts to join, to create better alignment with the state's established professional development regions; two of the three accepted. The partnership was ultimately dubbed "TEAM-Math," short for Transforming East Alabama Mathematics.

TEAM-Math's goal is improving grades K–12 mathematics teaching and learning in east Alabama, in particular focusing on improving mathematics achievement and closing achievement gaps among demographic groups. The partnership's mission statement (TEAM-Math 2009a), which was developed in 2003 by representatives of all the partners, follows:

> To enable all students to understand, utilize, communicate, and appreciate mathematics as a tool in everyday situations in order to become life-long learners and productive citizens by Transforming East Alabama Mathematics (TEAM-Math).

We have kept this mission statement at the center of the TEAM-Math effort, and we have displayed and discussed it at the beginning of all TEAM-Math events.

In November 2002, the leadership team recognized the potential match between its proposed directions and the Math and Science Partnership (MSP) program at the National Science Foundation (NSF). A proposal submitted to the MSP program in January 2003 was funded in October 2003, for $8.9 million, with a $400,000 supplement in 2005 to support the three additional districts. Internal funding from Auburn University ($100,000) allowed the partnership to begin activities in early 2003, prior to any response from NSF.

Although the large-scale funding from NSF allowed TEAM-Math's activities to ramp up very quickly, the partnership has maintained the stance that receiving the large grant was only a means to an end, and that we would seek other sources of support for activities that support the mission. That additional funding has come from several sources. The Malone Family Foundation contributed $305,000 to enhance the use of technology among secondary school mathematics teachers. Grants of $600,000 and $1,500,000 from the NSF Noyce Scholarship Program supported developing the TEAM-Math Teacher Leader Academy. Master teacher fellows in the Academy receive an annual stipend, tuition reimbursement, and professional development focusing on leadership development. NSF contributed a final supplement of $200,000 to support the partnership through September 2011.

The remainder of this chapter is organized into six sections. The first section provides the research foundation that guided our enterprise. The second section describes the organizational arrangements that support our research, policy, and practice interactions. The next section gives insights into the policy-related factors that are central to our efforts. We then discuss how we evaluated the partnership and conclude with some of our unique features and outcomes, along with final insights.

TEAM-Math: The Research Foundation

What research was foundational to our enterprise? From its inception, lessons learned from related projects and research literature have guided TEAM-Math's thinking and design. A primary source of insight was the Mathematics: Application and Reasoning Skills (MARS) project (Campbell et al. 2003). The TEAM-Math codirector, Marilyn Strutchens, was a member of the MARS project team; we replicated a number of the MARS's components, including school-teacher leaders, the development of a curriculum guide, summer institutes, school administrator briefings, and parental involvement. We also used the cohort approach for schools to enter the project. What follows is a brief description of some of the major principles derived from lessons learned in the MARS project and other research.

Leadership

The partnership used a distributed leadership model (see Spillane [2000]) as a foundation for the organizational structure. We believe it important that leadership be shared, thus including the voices of teachers, administrators, mathematicians, mathematics educators, and evaluators. We have a committee for each major area of activity—professional development, teacher preparation, outreach, and evaluation. Each committee consists of at least two representatives from the grades K–12 partners (two teachers or a teacher and an administrator), at least two mathematicians, at least two mathematics educators, and graduate students. This structure has enabled us to keep the schools' needs and goals in mind as we advocate change in their schools and across the districts.

Professional Development

The foundations of the professional development that we have provided to teachers have been (1) research related to helping teachers assist students in reaching their full mathematics potential; (2) effort at raising teachers' awareness of students' needs, both academic and cultural; and

(3) effort at providing teachers with the knowledge to integrate assessment with teaching, so that they are constantly aware of students' progress and whether what is being taught is understood (Bransford, Brown, and Cocking 1999). One of the project's main goals is to produce an articulated school effort to enhance students' motivation and achievement by improving teachers' attitudes toward, and use of, reform practices (i.e., those consistent with the recommendations of *Principles and Standards for School Mathematics* (National Council of Teachers of Mathematics [NCTM] 2000). The practices are student-centered, and they provide instructional scaffolding for students that allows them to move from what they know to what they do not know (Carpenter et al. 1999; Carpenter, Franke, and Levi 2003; Ladson-Billings 1995). Teachers use a variety of ways for students to explore curriculum content—a wide selection of sense-making activities or processes through which students can come to understand and "own" information and ideas. Teachers also have many options available through which students can demonstrate or exhibit what they have learned (Haberman 1992; Senk and Thompson 2003; Tomlinson 1995).

Our professional development focuses on supporting school mathematics reform following the research and recommendations of Borasi and Fonzi (2002); Loucks-Horsley, Stiles, and Hewson (1996); and Loucks-Horsley et al. (2003). Thus, a clear, well-defined image of effective classroom learning and teaching drives the professional development. The professional development does the following:

- gives teachers opportunities to develop knowledge and skills and broaden their teaching approaches so they can create better learning opportunities for students;

- uses instructional methods that mirror the methods to be used with students to promote learning for adults;

- builds or strengthens the learning community of mathematics teachers;

- prepares and supports teachers to serve in leadership roles if they are inclined to do so;

- consciously provides links to other parts of the educational system; and

- includes continuous assessment (Loucks-Horsley, Stiles, and Hewson 1996).

Community and Parent Involvement

We used ideas from a number of sources that stressed the importance of working with and informing parents and other stakeholders—in timely, productive ways—about the advocated changes in mathematics teaching and learning (Meyer, Delagardelle, and Middleton 1996; Peressini 1998; Strutchens 2002; Hendrickson et al. 2004).

Overall, we have consistently tried to link research to practice reciprocally. Our goals have been basing our work with project constituents on research related to systemic teacher change and listening to constituents' voices as we make major decisions for the project. A consistent theme throughout the partnership has been examining what has already been done through reviews of related literature and then determining what can be adapted for our situation. We used the primary elements—developing teacher leaders, providing reform-based professional development, and taking the constituents' needs into consideration—to formulate the partnership's design and organization, which are described in the following section.

TEAM-Math's Organizational Design

What TEAM-Math organizational arrangements support our research, policy, and practice interactions? We developed a systemic change model to work toward the partnership's goals and support its mission. We derived the model from the research foundation discussed in the previous section and lessons learned from MARS. The different components of the model were developed to support mathematics teaching and learning by informing and supporting partnership stakeholders. What follows briefly describes the model for change strategy.

We undertook curriculum alignment to ensure that a common vision existed for what should be taught across the partnership. Representatives from each of the partners met to create a common curriculum guide aligned to state and national standards and incorporating research and best practices (TEAM-Math 2009b). The districts subsequently adopted this guide, in some instances with minor modifications. The guide continues to be updated annually. In addition, the partnership formed a textbook adoption committee to offer guidance to the partners. The committee recommended a coadoption model, which used a more-traditional textbook series in conjunction with one of the more investigative sets of materials developed with NSF support that were more in alignment with project goals—Investigations in Number, Data, and Space for grades K–5; Connected Math Project for grades 6–8; and the Interactive Mathematics Program for grades 9–12. We took this approach because we would implement the new textbooks before teachers would have time to participate in the training necessary for the NSF materials' successful implementation. All 12 original school district partners adopted the recommended materials. This common curriculum has provided a firm foundation for other partnership activities.

Teacher-leader development was another part of the model. This part of the change strategy consisted of professional development offered to teachers identified as school- and district-level teacher leaders, who were expected to serve as a liaison to the partnership for the school or district and to support the implementation of partnership goals. The teacher leaders worked with individual teachers to improve their skills, organized and conducted school-based planning and inquiry groups, and developed teacher-learning communities at their schools. With only a few exceptions, these teacher leaders continued to function as full-time classroom teachers. The partnership provided quarterly, half-day workshops for the teacher leaders, focusing on general development of leadership skills and specific roles they could carry out. Although having full-time teachers act as teacher leaders limits the extent of their activity, it also gains them some credibility with other teachers (Lord and Miller 2000). We arranged for substitutes for teacher leaders, to allow them to take some days to work with teachers in their classrooms. Teacher leaders also held after-school workshops or facilitated workshops during teachers' in-service training days at their schools.

Professional development was central to the change strategy model. We have provided professional development to teachers at 94 schools that joined one of five cohorts, beginning with summer 2004 through summer 2008. Schools applied for admission to a cohort on the basis of the levels of teachers' participation and administrative support. Teachers at accepted schools attended a two-week summer institute focusing on content knowledge, research-based pedagogical practices, and grade-level sessions exploring the recommended instructional materials along with school and district planning. The teachers returned for a one-week institute the following summer that expanded on the first summer's themes and helped them continue to develop their understanding of how to support better the mathematical learning of all students. In addition, the project expected teachers to attend half-day, quarterly meetings held on Saturday mornings; these meetings built on the summer institutes' themes and focused on the material from the curriculum guide to be taught in the coming nine weeks.

We consider engagement with all stakeholders an essential part of the model. Stakeholder involvement focused on increasing the awareness of and support for TEAM-Math. General information related to the partnership is on the Web site, www.TEAM-Math.net. Periodic mailings and e-mails provided updates on upcoming activities. School administrators and guidance counselors attended special workshops held during the summer institute and follow-up workshops during the school year. These workshops instilled a general understanding of mathematics teaching and learning as well as information on practices that better supported the partnership's goals. For example, when visiting a classroom, an administrator should have expected to see students communicating about mathematics, not just quietly working on worksheets or taking notes from the teacher's lecture.

Several activities promoted parental and community involvement in students' mathematics education. The partnership initially held partnership-wide parental briefings in a central location. This proved less than optimally effective, so the partnership instead encouraged the teacher leaders to organize school-based parental involvement events. The Multicultural Literature as a Context for Mathematical Problem Solving program (Strutchens 2002), in which teachers received professional development training to become facilitators of family learning events, has also been implemented across the partnership. After completing the training, the teachers held six weekly sessions in which parents and children explored mathematical problems together. Each session was based on a children's book that has a multicultural theme and embedded mathematics or has the potential to evoke mathematical discussion.

Teachers' preparation is the project's final major component. The universities have worked to improve their preparation of new teachers, revising the mathematical content that preservice teachers receive as needed. The universities also developed a new, "capstone" course for secondary mathematics education majors that supports learning the skills and understandings associated with making connections between higher mathematics and the school curriculum. Further, the partnership has made a specific effort to improve preservice teachers' field experiences by placing them with teachers active in the TEAM-Math partnership.

The components of the TEAM-Math model are interdependent, as depicted in figure 9.1. Changes across the educational system must be made in order to provide a climate in which students receive a substantively different experience in their mathematics classroom, hypothesized to lead to improving students' outcomes and narrowing gaps in mathematics performance among demographic groups. TEAM-Math was organized with this hypothesis in mind. Figure 9.1 reflects our theory of action.

Partnership and Policy

What policy-related factors are central to our partnership? The partnership's systemic approach includes significant focus on policy issues. Perhaps the most notable is the curriculum alignment component, which created a common vision for mathematical content on which all other

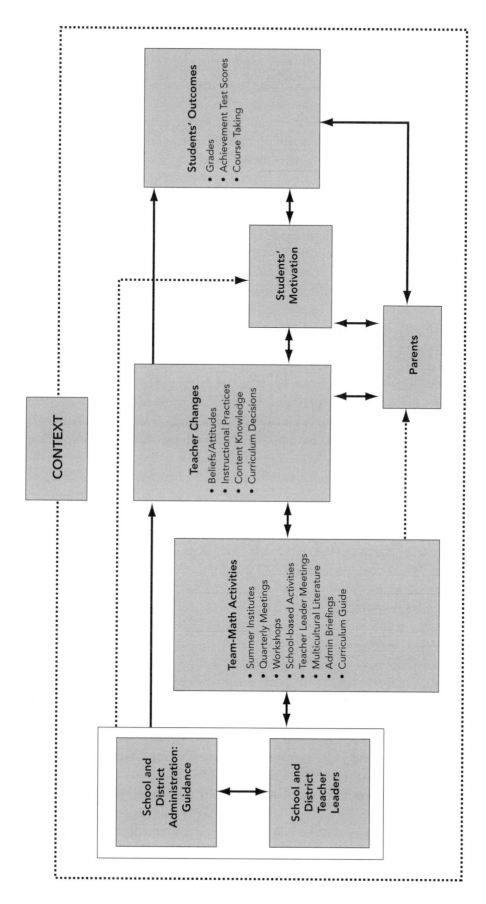

Fig. 9.1. TEAM-Math logic model

aspects of the partnership were built. For example, the professional development focused more on the common curriculum guide and instructional materials, the summer institutes offered the curriculum's general vision, and the quarterly meetings helped teachers prepare to implement teaching approaches that the project advocated through the curriculum in the next nine weeks. Likewise, the teacher leaders easily collaborated across schools as they sought ways to improve instructional practices.

Other areas of policy engagement were more challenging. Administrators face a wide range of often-conflicting initiatives coming from both the state and the district, and we worked hard to ensure that their commitment to the partnership went beyond verbal assent to implementing policies and practices in alignment with the partnership goals. Even so, we encountered several major challenges. As the partnership began, the state made a major commitment to the Alabama Reading Initiative (ARI) (ALSDE 2009c). In many instances, administrators faced a conflict in priorities between investing time in ARI or TEAM-Math. Since ARI was a state program, many felt an obligation to put that first. We actively encouraged schools and districts to develop a plan for participation in both programs that would not unnecessarily delay schools' participation in one of the TEAM-Math cohorts.

Along with Alabama's emphasis on reading came the challenge of ensuring policies that provided adequate instructional time for mathematics. Schools involved in ARI allocated up to three hours of reading instruction with minimal attention to other areas. At one point, the state assistant superintendent of instruction put out a memo appearing to suggest that instruction in areas other than reading should be considered secondary. Through our interactions with the administrators and teacher leaders, we tried to reemphasize the importance of mathematics as a foundational area of focus for the elementary grades.

Finally, the focus on Adequate Yearly Progress mandated through No Child Left Behind led many administrators to focus more on actions that might quickly boost test scores than on quality instruction in mathematics. We tried to demonstrate that the two were not mutually exclusive and that an emphasis on quality mathematics instruction would lead to improvements in test scores, whereas efforts to inflate test scores artificially would not create the foundation of mathematical learning needed for long-term improvement.

To address these challenges, the partnership engaged in dialogue about the issues directly with administrators at events designed for them. These events often included panel discussions led by administrators who were effectively using district or school policies to support project goals. For example, one district administrator personally wrote each teacher a letter stating that (1) their participation in the partnership professional development was expected and (2) those who did not participate would need to submit an explanation to him. Moreover, we addressed policy issues with the teachers during professional development events and made particular efforts to furnish information and tools to teacher leaders that they could use in working with their administrators. We also scheduled individual meetings with administrators who teacher leaders reported were implementing policies and practices that conflicted with the partnership's goals.

A final area of policy focus was helping administrators interpret state directives regarding teachers' evaluations in ways that aligned with the project goals. The state Professional Evaluation Personnel Evaluation System (ALSDE 2009a), which is now being revised, included rating items that led administrators to rate effective instructional practices as ineffective. For example, one item asked whether the teacher clearly identified the lesson's objectives at the start of the lesson. For teachers who were promoting inquiry into mathematical concepts, this could short-circuit the activities they had planned. Our professional development for administrators included attention to alternative interpretations of these and other objectives—for example, giving a nonspecific objective at the beginning of the lesson, but making sure to state a more-specific objective by the end of the lesson, could meet the item's intent.

Ensuring policy alignment has been a continuing struggle throughout the years of the partnership. As teachers encountered policies that they saw as misaligned with what the partnership promoted, they encountered either frustration, if they continued to pursue what they had come to see as best practices supporting their students' mathematical learning, or resignation, if they decided that the mixed messages suggested that there was no real support for change.

Evaluating the Partnership

How did TEAM-Math evaluate the partnership? Evaluation was a central component of the partnership. We collected data from a wide range of stakeholders engaged in the partnership. Our ultimate goal was to affect students' mathematics achievement. We relied on data from state-mandated assessments—the Stanford Achievement Test, the Alabama Reading and Mathematics Test, and the Alabama High School Graduation Exam—to measure students' progress. In addition, students completed a survey that gave information about their attitudes toward learning mathematics, which is related to achievement, as well as their impressions about what happens in the mathematics classroom.

Teachers also received a survey, which gave us information about their attitudes about mathematics and teaching mathematics, as well as their impressions of what happens in their classrooms. We then linked the data from students and teachers, in order to triangulate data related to classroom practices. The other major data source from teachers was the numbers of hours they participated in TEAM-Math professional development, which gave a general indication of their involvement in the partnership.

We also administer surveys to administrators and parents. The administrators' survey gathered information on their beliefs about mathematics teaching and learning, their view of TEAM-Math's effectiveness at their school, and their administrative practices related to TEAM-Math. The parents' survey gathered information about parents' beliefs about mathematics teaching, their views of mathematics instruction at their children's school, and their involvement in their children's mathematics education, including in TEAM-Math-related parent events. These measures linked to others to paint a more reliable picture of how a system's various parts relate to one another, comparing different stakeholders' views on how mathematics teaching and learning was changing.

In addition, the evaluation strategy included a qualitative component involving site visits to schools selected to represent different levels of implementation as well as grade levels. These site visits included interviews with teachers, teacher leaders, administrators, parents, and students.

We used evaluation instruments, such as surveys of students and teachers and achievement data from state assessments, to keep the partners informed of our progress. Analyses of the data gathered from these instruments and data sources help teachers and administrators see the importance of the changes being made. This was particularly important early in the partnership, to ensure continued buy-in as we could demonstrate initial positive changes. Even though we kept the individual schools and teachers anonymous, school partners got a sense of the impact of the project on their students' achievement and attitudes.

We also evaluated each major event that we conducted in order to determine if we were meeting participants' needs and maintaining our professional development's quality. Although the evaluations have generally been very positive, we also pinpointed weak areas—or presenters who were having problems getting their messages across to participants—and subsequently addressed those issues.

Unique Features and Outcomes

What unique features and outcomes have come from the TEAM-Math partnership? One of the project's major features that has been essential to professional development and increased the partnership's cohesiveness is the curriculum guide and its effect on common textbook adoption. The curriculum guide was based on the Alabama Course of Study (ALSDE 2003), state (ALSDE 1999, 2005) and national (National Assessment Governing Board 2002, 2008) assessment frameworks, and *Principles and Standards for School Mathematics* (NCTM 2000). A TEAM-Math committee was charged with building a coherent, comprehensive, user-friendly reference. The curriculum guide, along with the recommendations for the partnership from the textbook adoption committee, led to unanimous adoption of the same textbooks across the grade bands in all the districts. We have continually upgraded the curriculum guide to include references to the textbooks, pacing guides, and other details that have made it invaluable to teachers.

The project's teacher leader component has also been very strong. In fact, the East Alabama Council of Mathematics owes its existence to teachers wanting a way to continue networking after TEAM-Math ends. Teacher leaders interacted with one another beyond the formal meetings and workshops that we organized for them. Many visited TEAM-Math colleagues in classroom settings. Moreover, many of our teacher leaders served as presenters for the project's professional development events and for the Alabama Mathematics, Science, and Technology Initiative (AMSTI), a statewide program addressing grades K–12 and funded by Alabama's MSP program and the state budget. This initiative aligned with TEAM-Math's approach to mathematics education. In addition to professional development, AMSTI offered instructional materials in the form of kits, designed to help the teachers implement targeted units, and classroom support in the form of AMSTI mathematics specialists who work with individual teachers. AMSTI activities complemented those of TEAM-Math very well: the local AMSTI site worked closely with TEAM-Math to ensure coordination. To date, 43 schools that have participated in TEAM-Math have transitioned to AMSTI. Although data on this project's impact on schools in east Alabama is not yet available, AMSTI has reported significant changes in students' achievement across the state. A number of the teacher leaders have gone on to take more formal school, district, and state leadership positions. Focus groups and other means have shown us that teacher leaders are essential to AMSTI's successful implementation in their schools.

We made efforts to ensure that partnership participants had quality professional development available. Our professional development's associated quality factors focused on pedagogical content knowledge, curriculum materials knowledge, vertical alignment, content knowledge, and equity issues. We have also tried to have teachers experience mathematics teaching and learning the way that we would like to have their students experience it. Linking research to practice has been a priority.

The TEAM-Math project's parental component has also been a success, especially in elementary and middle schools. Through family math nights, briefings, and the Multicultural Literature as a Context for Mathematical Problem Solving with Parents and Children Learning Together program (MCL for short), teacher leaders and administrators could successfully help parents understand the changes in mathematics classes. Over a four-year period, 2004–08, 29 schools implemented the MCL program, with each school averaging a total of about 85 participants including children and their parents. Moreover, some schools have reported more than 300 parents and children attending MCL sessions or other family math night events. To date, we have had virtually no major negative responses to the partnership from parents or other community stakeholders.

The seminars that we have held for mathematics educators, mathematicians, and our respective graduate students have also been helpful. These seminars have helped us get on the same page concerning teachers' mathematical education. When we began the project we often talked past one another. Now we can converse more productively. Through the seminars, we are also conceptualizing papers related to the project's other various aspects.

The TEAM-Math Annual Conference on the Mathematical Education of Teachers, held at Tuskegee University's Kellogg Conference Center, has also been very beneficial to the partnership. This national conference has the distinctive feature of taking place in a very rural area with a large number of high-needs schools, as opposed to most other conferences, which are held in urban settings. Teachers, many from these rural schools and who otherwise would not have been able to attend a conference, have had opportunities to hear from mathematics educators and mathematicians about whom they may have read or whose work they may have studied. Moreover, participants have been able to interact with one another in ways that may not have been possible in the past.

Finally, we have had some positive outcomes for students, and we have begun to discover crucial factors leading to those outcomes. Analyses of survey and achievement data (e.g., Martin, Strutchens, and Karabeneck [2009]) have demonstrated that students who reported teachers' greater use of reform practices, teachers' higher expectations, and teachers' higher standards showed more desirable levels of motivation to learn mathematics. Further, teachers' use of reform practices and higher expectations had both direct effects on achievement test scores and indirect ones mediated by students' motivation. Other analyses have shown a relationship among teachers' participation in comprehensive professional development, their beliefs about reform practices, and their use of those practices in their classrooms (Martin et al., in press). This relationship thus establishes a link between the professional development teachers have received and their students' attitudes and performance. Qualitative analyses (Strutchens, Henry, and Martin 2009) have shown that school-level support is an important factor in successfully implementing TEAM-Math practices. Schools with effective administrator and teacher leadership are more likely to have teachers enthusiastically involved in the partnership and students aware of the changes in their teachers' instruction.

Final Thoughts

Our central recommendation is to develop and nurture partnerships at all levels. To a large extent, success builds on effective relationships. Treat teachers and administrators as partners, value their expertise, make available the necessary support, and keep the project goals on their minds. We have found that being visible to, available to, and appreciative of pivotal school personnel is important to the project's work, especially since so many other initiatives can consume their attention. Administrators' active support is extremely important, but so is that of teacher leaders who can serve as role models and encourage their colleagues to participate. Some teachers will implement changes because they believe in them, but many will wait to see if support for those changes really exists. One teacher leader expressed this well: he stated that many people had come to east Alabama with promises of what they could provide and left before anything of substance happened. TEAM-Math's leaders, however, had stayed and made substantial contributions to mathematics teachers' growth in the area. The importance of perseverance cannot be underestimated.

Teachers must have central roles in developing project products and feel ownership toward them. The curriculum guide and the textbook adoption review report were primarily developed by teachers, thus teachers were able to defend the documents to any naysayers. Many of the teacher leaders feel deeply invested in the partnership and will rise to its defense.

It is also crucial that disciplinary faculty get involved in activities that help them to understand the value of pedagogical content knowledge and effective curriculum materials. Having the support of both mathematicians and mathematics educators adds credibility to the recommendations being made. In addition, mathematicians involved in TEAM-Math have reported that they have personally adopted the teaching methodologies and resources advocated by TEAM-Math, with the result that their students have developed a better appreciation for and a deeper understanding of course content. These findings should serve as a source of motivation for all science, technology, engineering, and mathematics faculty to seriously consider instructional strategies other than the traditional lecture style of teaching.

We have learned, too, not to underestimate teachers. Although some of the partners questioned whether teachers would engage in reading mathematics education research, we have found that teachers will respond in positive ways to professional development that is based on national and state standards and contains best practices. Finding teachers who believe in the project and have implemented the curriculum to serve as presenters is also essential to the project's success. They increase credibility among participants and encourage continued participation.

After more than six years of operation, the product of the TEAM-Math partnership is a greener landscape in terms of mathematics teaching and learning in east Alabama, thanks to the concerted efforts by a comprehensive group consisting of hundreds of dedicated teachers, administrators, mathematics teacher educators, mathematicians, and graduate and undergraduate students. Many of these people individually desired to improve mathematics education prior to the inception of TEAM-Math, but no one person working in isolation could possibly bring about the systemic change that has occurred. We needed a comprehensive partnership to bring stakeholders together to identify, understand, and address all the issues facing mathematics education in east Alabama. The name the partnership adopted in 2003—"TEAM-Math," for "Transforming East Alabama Mathematics," reflects this idea, with the emphasis on "transforming," not just making minor adjustments. Moreover, the abbreviation evokes the sense of community and common purpose that we were then beginning to feel and that has grown over the years.

REFERENCES

Alabama State Department of Education (ALSDE). *Mathematics Specifications for the Alabama High School Graduation Exam.* Montgomery, Ala.: ALSDE, 1999.

——— *Alabama Course of Study: Mathematics.* Montgomery, Ala.: ALSDE, 2003.

———. *Alabama Reading and Mathematics Test: Specifications for Mathematics.* Montgomery, Ala.: ALSDE, 2005.

———. "Alabama Professional Education Personnel Evaluation Program." http://www.alabamapepe.com, 2009a.

———. "Alabama Professional Education Personnel Evaluation Program." http://www.alabamapepe.com, 2009b.

——— "Alabama Reading Initiative." http://www.alsde.edu/html/sections/section_detail.asp?section=50, 2009c.

Borasi, Raffaella, and Judith Fonzi. *Professional Development That Supports School Mathematics Reform.* Foundations Monograph No. 3. Arlington, Va.: National Science Foundation, 2002.

Bransford, John D., Ann L. Brown, and Rodney R. Cocking, eds. *How People Learn: Brain, Mind, Experience, and School.* Washington, D.C.: National Academies Press, 1999.

Campbell, Patricia F. "Elementary Mathematics Specialists: A Merger of Policy, Practice, and Research." In *Disrupting Tradition: Research and Practice Pathways in Mathematics Education,* edited by William F. Tate, Karen D. King, and Celia Rousseau Anderson, pp. 93–103. Reston, Va.: National Council of Teachers of Mathematics, 2010.

Campbell, Patricia, Andrea Bowden, Steve Kramer, and Mary Yakimowski. *Mathematics and Reasoning Skills: Final Report (Revised).* Grant No. ESI 9554186. College Park, Md.: University of Maryland, MARS Project, 2003.

Carpenter, Thomas P., Elizabeth Fennema, Megan Loef Franke, Susan Empson, and Linda Levi. *Children's Mathematics: Cognitively Guided Instruction.* Portsmouth, N.H.: Heinemann, 1999.

Carpenter, Thomas P., Megan Loef Franke, and Linda Levi. *Thinking Mathematically: Integrating Arithmetic and Algebra in Elementary School.* Portsmouth, N.H.: Heinemann, 2003.

Haberman, Martin. "The Pedagogy of Poverty versus Good Teaching." *Education Digest* 58 (September 1992), 16–20.

Hendrickson, Scott, Daniel Siebert, Stephanie Z. Smith, Heidi Kunzler, and Sharon Christensen. "Addressing Parents' Concerns about Mathematics Reform." *Teaching Children Mathematics* 11 (August 2004): 18–23.

Ladson-Billings, Gloria. "Making Mathematics Meaningful in Multicultural Contexts." In *New Directions for Equity in Mathematics Education,* edited by Walter G. Secada, Elizabeth Fennema, and Lisa Byrd Adjian, pp. 126–45. New York: Cambridge University Press, 1995.

Lord, Brian, and Barbara Miller. *Teacher Leadership: An Appealing and Inescapable Force in School Reform?* Newton, Mass.: Education Development Center, Inc., 2000.

Loucks-Horsley, Susan, Katherine Stiles, and Peter W. Hewson. "Principles of Effective Professional Development for Mathematics and Science Education: A Synthesis of Standards. *NISE Brief* 1, no. 1. Madison: University of Wisconsin—Madison, National Institute for Science Education, 1996.

Loucks-Horsley, Susan, Nancy Love, Katherine E. Stiles, Susan Mundry, and Peter W. Hewson. *Designing Professional Development for Teachers of Science and Mathematics.* 2nd ed. Thousand Oaks, Calif.: Corwin Press, 2003.

Martin, W. Gary, Marilyn E. Strutchens, and Stuart A. Karabenick. "Changing Teachers' Attitudes and Practices through Professional Development." Math and Science Partnership Learning Network Conference, Washington, D.C., 2009. http://hub.mspnet.org/index.cfm/msp_conf_2009_abstracts.

Martin, W. Gary, Marilyn E. Strutchens, Michael A. Woolley, and Melissa Gilbert. "Transforming Mathematics: Teachers' Attitudes and Practices through Intensive Professional Development." In *Motivation and Disposition: Pathways to Learning Mathematics*, Seventy-third Yearbook of the National Council of Teachers of Mathematics (NCTM), edited by Daniel J. Brahier, pp. 291–303. Reston, Va.: NCTM, in press.

Meyer, Margaret R., Mary L. Delagardelle, and James A. Middleton. "Addressing Parents' Concerns over Curriculum Reform." *Educational Leadership* 53 (April 1996): 54–57.

National Assessment Governing Board (NAGB). *Mathematics Framework for the 2003 National Assessment of Educational Progress.* Washington, D.C.: NAGB, 2002.

———. *Mathematics Framework for the 2009 National Assessment of Educational Progress.* Washington, D.C.: NAGB, 2008.

National Center for Education Statistics. "State Assessments." http://nces.ed.gov/nationsreportcard/mathematics/stateassessment.asp, 2009.

National Council of Teachers of Mathematics (NCTM). *Principles and Standards for School Mathematics.* Reston, Va.: NCTM, 2000.

Peressini, Dominic D. "The Portrayal of Parents in the School Mathematics Reform Literature: Locating the Context for Parental Involvement." *Journal for Research in Mathematics Education* 29 (November 1998): 555–82.

Senk, Sharon, and Denisse Thompson, eds. *Standards-Oriented School Mathematics Curricula: What Does Research Say about Student Outcomes?* Mahwah, N.J.: Lawrence Erlbaum Associates, 2003.

Spillane, James P. "Cognition and Policy Implementation: District Policymakers and the Reform of Mathematics Education." *Cognition and Instruction* 18 (2) (2000): 141–79.

Strutchens, Marilyn E. "Multicultural Literature as a Context for Problem Solving: Children and Parents Learning Together." *Teaching Children Mathematics* 8 (April 2002): 448–55.

Strutchens, Marilyn E., Daniel Henry, and W. Gary Martin. "Improving Mathematics Teaching and Learning through School-based Support: Champions or Naysayers." Math and Science Partnership (MSP) Learning Network Conference, Washington, D.C., January 2009. http://hub.mspnet.org/index.cfm/msp_conf_2009_abstracts.

TEAM-Math. *2003 TEAM-Math Curriculum Guide.* Auburn, Ala.: TEAM-Math, 2003.

———. "Mission Statement." Auburn, Ala.: TEAM-Math, 2009a. http://TEAM-Math.net/index.htm.

———. "TEAM-Math Curriculum Guide." Auburn, Ala.: TEAM-Math, 2009b. http://TEAM-Math.net/curriculum/index.htm.

Tomlinson, Carol A. "Differentiating Instruction for Advanced Learners in the Mixed-Ability Middle School Classroom." *ERIC Digest* E536. Reston, Va.: ERIC Clearinghouse on Disabilities and Gifted Education, Council for Exceptional Children, 1995.

CHAPTER

10

The SCALE Project: Field Notes on a Mathematics Reform Effort

Terrence Millar
Mathew D. Felton

I N THIS chapter we reflect on more than a decade of collaborative work at the University of Wisconsin (UW)—Madison and between the university and the Madison Metropolitan School District (MMSD) aimed at improving grades K–16 mathematics and science education. This chapter will focus on the mathematics initiatives, even though the science initiatives had some impact on those in mathematics.

This collaboration between UW and MMSD has received support from institutional resources and more than fourteen years of federal funding for projects awarded to UW. Examples of such funding include the Kindergarten through Infinity grant (KTI; funded twice by the National Science Foundation [NSF] GK–12 program, 1999–2001 and 2002–05); the National Institute for Science Education (NISE, an NSF "think tank," 1995–2000); the NSF comprehensive Math Science Partnership (MSP) "Systemwide Change for All Learners and Educators" project (SCALE, 2003 to present); the Center for the Integration of Research, Teaching, and Learning (CIRTL, 2003 to present) and Diversity in Mathematics Education (DiME, 1999–2004), both part of the NSF network of Centers for Learning and Teaching; and various U.S. Department of Education MSP Title IIB awards to MMSD administered through Wisconsin's Department of Public Instruction (DPI). Some of these initiatives have involved not only UW and MMSD, but also other partner universities and school districts from across the country, including California State University, Dominguez Hills (CSUDH), Denver Public Schools, and Los Angeles Unified School District (LAUSD). The first author was an interim codirector of NISE, and then principal investigator (PI) and project director, first for KTI and then for SCALE. The second author participated in SCALE as an undergraduate and graduate research assistant.

We have organized our arguments into four major sections. In the first section, we briefly discuss the research and evaluation framing that has guided our partnership effort. The second section describes more specifically the work of SCALE at the Madison, Wisconsin, site. The third section offers some insight into the cultural and institutional challenges associated with multiorganizational partnerships as well as in-house institutional dilemmas. The final section is a call for a systems approach to mathematics reform.

Research and Formative Evaluation Frame

All the federally funded initiatives at UW had research and formative evaluation as part of their structure. For example, in SCALE about 15 percent of the budget went to support the Research and Evaluation Team (RET). The RET took a multipronged approach, summarized in figure 10.1. Although the SCALE RET conducted no randomized studies, an associated research study, funded by NSF and led by the Wisconsin Center for Educational Research (WCER) director, is under way. This associated study involves randomized assignments of students in SCALE interventions in elementary school science in LAUSD. SCALE collected information directly at the institutional and partnership level, at the district level through case studies of policies and implementation affecting and affected by SCALE interventions, at the classroom level by observing SCALE implementation, and at the student level through an analysis of students' standardized test scores.

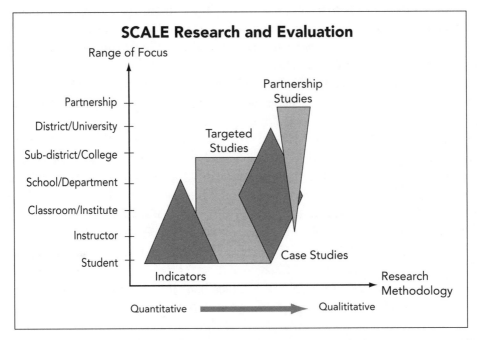

Fig. 10.1. Diagram of SCALE research approach

The overall impact of SCALE work on students' test scores was, for a variety of reasons, either unmeasurable or disappointingly small. If we give SCALE partners and leaders the benefit of the doubt concerning competence and vision, one conclusion is simply to acknowledge, once again, the challenges of retooling learning environments in complex, fluid, "noisy," and arguably underfunded environments. Although SCALE failed to deliver as hoped at the students' performance level, it achieved some positive effects and results at other levels of the system that arguably should be part of any comprehensive attempt to solve these challenges. We will revisit these results at the end of the chapter.

A great deal of this work also has drawn on, and adapted from, the existing research base to deal with pressing matters of practice, such as engaging teachers in meaningful professional development, improving preservice math and science content courses, and fostering greater collaboration within and among institutional entities (Century and Levy 2003; Darling-Hammond and Sykes 1999; Hall and Hord 2006; Loucks-Horsley et al. 2003; Senge 1990). Here SCALE and the work that preceded it benefited from not only the national research literature on mathematics education, but also previous UW/MMSD collaborative experiences and input from UW education faculty and emeritus faculty who are and have been quite prominent in this field.

SCALE at UW—Madison

The SCALE project sought to promote a cultural shift in its partner institutes for higher education through the use of collaborative teams. Doing this brought together differing forms of expertise by working across the institutional and disciplinary boundaries of school districts; colleges of education; and science, technology, engineering, and mathematics (STEM) disciplinary fields. This collaborative strategy grew out of the research literature and from previous experiences and lessons learned in such work as KTI. The goals of this cross-institutional collaboration were to improve preservice and in-service teacher preparation and learning, and to provide greater involvement and integration at the university level of faculty engaging in preservice and in-service activities. With respect to mathematics education in Madison, this meant collaboratively engaging the following partners.

- District instructional leaders and mathematics resource teachers

- Mathematics faculty and graduate students, particularly those involved in teaching content courses for preservice grades K–8 teachers

- Other STEM faculty interested in or willing to support in-service teachers' understanding of mathematics

- School of Education faculty and graduate students, particularly those with interest and expertise in mathematics education.

As will be seen below, SCALE supported collaboration among these partners in a number of ways. Two crucial methods existed, however, for supporting cross-institutional collaboration. First, SCALE drew on established relationships developed through prior SCALE work or past UW/MMSD partnerships to solicit engagement with new projects. In many ways, SCALE served as a means of increasing the coherence among the various projects and interests in grades K–16 mathematics education that had been occurring across UW and MMSD. A central facilitating factor, especially to engaging mathematics faculty and staff, was the role of the first author as the SCALE PI and as a member of the mathematics department. Second, SCALE and related grants set up financial support for faculty, graduate students, and district representatives. Financial support for faculty came in the form of summer salary or course releases that were particularly important for junior faculty. The role of financial support for faculty to focus on teachers' preparation is particularly relevant considering Hora and Millar's (2008) finding that UW participants in SCALE activities were concerned with the faculty workload at UW, which "gives relatively low priority to teaching excellence" (p. 57).

Only a very small percent of the SCALE award's total budget went to faculty and graduate students. A much larger percent funded teachers' stipends and release time so that those teachers could participate in SCALE professional learning activities. The funding's use, however, varied considerably across the four SCALE school districts, the five participating universities, and across the disciplines. For example, in LAUSD SCALE funds helped support both CSUDH science faculty and LAUSD science leaders as they collaborated to develop and deliver professional learning to LAUSD teachers. We should note that the LAUSD part of this work relied on a Quality Educator Development grant from the U.S. Department of Education. Such reliance on an additional grant highlights the budgetary challenges of institutionalizing educational change.

Major SCALE Activities Focusing on Mathematics Education

In this section we briefly detail three major activities supported by SCALE in Madison: the Math/Ed Liaison Committee, the Math Masters professional development program, and the MMSD Math Task Force. We discuss the accomplishments and the forms of support SCALE provided.

Math/Education Liaison Committee

One of SCALE's major accomplishments in Madison was achieving greater collaboration among the district, mathematics faculty and graduate students, and education faculty and graduate students surrounding the mathematics content courses for preservice grades K–8 teachers. SCALE, the school mathematics department, and school of education's leadership accomplished this by helping reconstitute the math department's formerly long-standing Preservice Committee. An ad hoc group made up of an associate dean from the School of Education, a former chair of the mathematics department, and the first author was central to reconstituting the committee with new leadership and to seeking out representation from MMSD and newly hired faculty in mathematics education. The committee moved forward on a common concern: that preservice grades K–8 teachers at UW received inadequate mathematical preparation for teaching middle school mathematics.

SCALE funding was a source of support for the faculty, especially in mathematics education, to participate in the activities related to middle school teachers' preparation. Moreover, SCALE funding was essential to the graduate students' work of actualizing committee goals in day-to-day practice, such as background research directed by a mathematics education faculty member, designing course lessons, and delivering course instruction. With SCALE's support, the Math/Ed Liaison Committee increased participation in a committee that many in the past had seen as largely ineffective and insular, developed new course syllabi more closely connected to grades K–8 mathematics teaching, created a new course sequence, and engaged the various partners in greater collaboration. The grades K–8 math content courses are typically taught by mathematics faculty, staff, and graduate students. One notable form of collaboration included frequent observation and collaboration on course instruction by district experts and education faculty who worked with some of the instructors of these courses, which led to some reported shifts in instructional practice (Hora and Millar 2008, p. 33). In recruiting faculty to participate, the first author was particularly careful to choose faculty with reasonable interpersonal skills and who did not hold, or appear to hold, district experts in "intellectual contempt." Even with this extra care, tension over participants' intellectual authority in both mathematics content and pedagogy marked many meetings.

Math Masters Professional Development

Another central collaborative effort supported by SCALE and U.S. Department of Education MSP Title IIB grants, dubbed the "Math Masters project," offered practicing middle school teachers mathematics content courses co-designed and co-taught by district mathematics resource teachers and UW STEM faculty. The MMSD teaching and learning director first approached the SCALE PI about this possibility. The director recognized the need for more mathematics content knowledge for middle school teachers, especially with the reform approach the district was advocating with its adoption of *Connected Math* (Lappan et al. 2008) as the middle school curriculum. The SCALE PI embraced this opportunity and that of engaging mathematics faculty in more complex pedagogical approaches than one normally found in university mathematics classes.

The Math Masters project had several important outcomes, including the participation of 438 teachers, among them teachers from surrounding districts, in more than fifteen workshops, statistically significant gains in teachers' content knowledge, greater collaboration and collegiality among UW faculty and MMSD, and UW faculty's deeper understanding of the middle school teaching and learning process (Hora and Millar 2008, p. 38). SCALE itself did not have sufficient funding available for the Math Masters project, so members of the SCALE partnership submitted a Title IIB Department of Education MSP proposal to support this extension of SCALE's work. The collaboration among UW mathematics faculty, other faculty, and district experts was considered so valuable that the school district subsequently pursued and received funding for two additional Title IIB grants—Expanding Math Knowledge (EMK) and the Science Masters Institutes (SMI). Although seeking additional monies represents an interest in continued collaboration, the fact that the collaborators had to go to external sources indicates the ongoing challenge of institutionalizing change that we discuss below. We should remark, however, that the need to pursue external funding is also a feature of research initiatives in engineering and the natural sciences at research universities, which depend on a steady stream of external funding. EMK's structure resembled that of the Math Masters program, but it focused on practicing elementary school teachers. The SMI approach also resembled Math Masters, but focused on practicing middle school science teachers. A professor from the Department of Mechanical Engineering, who had cotaught courses in the Math Masters program, volunteered to recruit science faculty from across UW and to help organize and run the program.

MMSD Math Task Force

At a MMSD Board of Education meeting toward the end of 2006, the Board approved a motion to conduct a comprehensive, independent, neutral review and assessment of the district's grades K–12 math curriculum and related issues. As with many districts across the country over the past two decades (Wilson 2003), mathematics instruction and curricula choices in Madison and some of the surrounding communities have been contentious. The Board stipulated that the MMSD superintendent appoint a task force, which the Board would approve. The superintendent, after consulting with the SCALE PI, appointed a ten-person task force and arranged for staff support for it from district and SCALE personnel. Although most task force members—six UW—Madison faculty and researchers, including a former chair of the mathematics department and members of the Math/Ed Liaison Committee; a parent; and a teacher—were drawn from the Madison community, the superintendent, in consultation with the SCALE PI, selected the committee's cochairs from outside the Madison community, to ensure that the review was independent and neutral. The appointed task force cochairs were a professor and former chair of the mathematics department at the University of Nebraska—Lincoln, who has played a national role in mathematics education; and a former Los Angeles Unified School District Deputy Superintendent of Instruction, now a faculty member in the department of educational leadership at California State University, Northridge, and a Graduate School of Education and Information Studies liaison at the University of California, Los Angeles.

The task force functioned as a learning community that met and communicated over a twelve-month period. This is important, since the mathematical, cognitive, educational, cultural, political, financial, and psychological issues raised by the Board of Education charge to the task force constituted a complex landscape. Although research and experience helped shed some light on this landscape, much was not—and still is not—understood. The extensively researched *Final Report of the Task Force* included a set of findings and recommendations. For example:

> Finding 1: The single most important step that the MMSD Board of Education can take in support of students' improved achievement in mathematics is to align district goals, policies, and resources in ways that result in a mathematics teacher workforce well prepared in mathematics content and in techniques for teaching mathematics. This issue is especially crucial in grades 5–8.

> Finding 5: The district's curriculum should develop conceptual understanding, computational fluency, and problem-solving skills simultaneously. Debates on the relative importance of these aspects of mathematical knowledge are generally misguided.

MMSD changed superintendents before the task force delivered its report to the Board. The new superintendent, however, embraced the report's findings and recommendations and has begun taking collaborative steps with UW administration to act on them.

Madison: A Cultural Challenge

SCALE and the work leading up to SCALE had support at the highest levels in both UW and MMSD. The PI and project director is a mathematics professor and associate dean for the physical sciences in the Graduate School at UW, which is also the university research office. He assumed his KTI and SCALE roles as part of his responsibilities in both appointments. For example, UW, to support this work funding, has paid 25 percent of the first author's salary for more than a decade. As another example, KTI led to creating a new office in the Graduate

School for graduate students' professional development. During SCALE, the PI met often with the deans of Education and Letters and Sciences, as well as the MMSD superintendent and instructional leaders. These meetings collaboratively managed such aspects of the project as arranging support for the SCALE research, aligning teachers' professional learning initiatives with state and district standards and resources, and affecting policy to improve the overall educational system surrounding math and science education. Because of complex cultural issues, and despite this level of support and collaboration at the administrative level, less coherence existed at the UW departmental level. We believe these cultural issues are an important part of the "invisible" challenge of the last two decades to national efforts to collaborate across disciplines and types of institutions to improve mathematics education. For example, the SCALE RET's research showed the need for a purposeful, informed approach to engaging the cultural practices of STEM departments (Hora 2008, p. iii):

> The enduring lesson from SCALE activities at UW—Madison is that efforts to change the culture of teaching and learning in STEM departments should focus on illuminating and then shifting the pervasive cultural schemas that faculty hold for teaching and learning. One strategy for doing so is to create officially sanctioned venues where individuals from different disciplinary backgrounds are led by a skilled facilitator or "culture broker" to focus on commonly shared pedagogy-related challenges. Leaders would benefit from being aware of and sensitive to the deeply entrenched nature of cultural schemas and their embeddedness in the local institution.

These departmental and disciplinary cultural issues are alive and well, and they add to the challenge of inclusive collaborative partnerships and institutionalized shifts in practice.

The Conundrum of Systemic Change

We have found that traditional cultural practices and lines of authority in mathematics reform are often treated as invariants instead of parts of the challenge that need to be addressed. To highlight our concerns, we offer personal perspectives on the challenges of systemic change.

> As an undergraduate mathematics major and later an education graduate student working as a SCALE research assistant, I was able to participate in a number of collaborative activities with the mathematics department and the school district. These strengthened my own understanding of pre- and in-service teacher preparation and supported me in learning to work collaboratively to integrate my and others' differing forms of expertise. I designed courses with mathematicians and mathematics resource teachers from the district, and benefited from their feedback when I taught sections of these courses. I also witnessed a number of similar relationships develop around me. Over multiple semesters mathematicians were working alongside district experts on engaging teachers in professional development and district experts were meeting with mathematicians to discuss the challenges of teaching preservice teachers in a meaningful way, often observing or co-teaching class. As far as I understood it this level of collaboration, particularly between the mathematics department and the district, was unprecedented in the Madison context.
>
> A concern I have as SCALE wraps up and I prepare to move on to a new institution is the ability of these changes to persist in the absence of key graduate students with a particular interest and expertise in this area and without the resources, particularly financial resources, needed to support continued engagement by graduate students and faculty. Despite the collaborative work and some improvements in the value individuals place on what different experts bring to the

table, the preservice K–8 math content courses are plagued by institutional constraints. There remains no institutionalized program for supporting the instructors (who are predominantly mathematics graduate students) either before or during the semester. Therefore, many instructors, especially in the future, may have little or no familiarity with the daily realities of K–8 mathematics teaching, and without the support provided by SCALE it is hard to imagine how instructors will develop this understanding.

The narrative above, of the second author's experiences with SCALE, captures one of the biggest challenges we have faced in our work: dealing with institutional inertia when attempting systemic change. The RET's case study of the effect of SCALE activities on UW has documented this difficulty. They found that although SCALE was successful in bringing about a number of localized changes (e.g., shifts in the views of individual faculty members, and greater collaboration between STEM and Education faculty), it did not result in a radical shift in institutional practice. The authors argue that effecting substantive change at UW—Madison will likely require a comprehensive strategy that matches the complexity of the institutional environment (Hora and Millar 2008, p. 64). Although the details of how this challenge manifests itself will vary from institution to institution, concern over it extends beyond our particular context. The Westat evaluation report (Zhang et al. 2009) on the National Science Foundation's Math Science Partnership program found that "few benefits extend beyond those faculty who are direct participants, and few systemic changes have been made in [institutions of higher education]" (p. 6-1).

A number of factors contribute to the challenge of effecting substantive institutional change. As the SCALE case study cited above alludes, complex cultural issues come in to play at the department and other unit levels, and at their interfaces. The first author reflects on this in the following narrative:

> Even as a mathematics professor with extended cross-cultural experiences (as a Marine in Vietnam and as an anthropologist's spouse in Indonesia) and significant administrative experience (as an associate dean) in promoting graduate education and research at a major research university, the complexities involved in the issues surrounding mathematics instruction still remain baffling to me.
>
> Often the problems are attributed to issues involving personalities or professional competence: "X speaks with such unconscious condescension that it is breathtaking," "Y does not have the mathematical talent to warrant trust in his educational opinions," "Z does not have the pedagogical skill necessary to be entrusted with future teachers' educations," etc. In my administrative experiences, negative ad hominem characterizations often mask structural flaws. That is, if a system has a structural weakness or is internally contradictory, oftentimes this creates enough stress in associated individuals that it brings out their worst. But often the causality involved is that the structural issues bring out the bad behavior, and not the other way around.
>
> In the case of mathematics learning, it just might be that hidden complexity in nominally "clear settings" creates this kind of "invitation" to negative ad hominem analysis. That is, the issues of productive rule-governed symbol manipulation in a conceptually tiered disciplinary area that has evolved over a couple of millennium, coupled with the characteristics of human learning and instructional practices and environments intended to promote understanding and proficiency, just might still be sufficiently beyond our abilities to understand that we more often than not concentrate on dividing weaknesses rather than on integrating strengths.

Why is it difficult to institutionalize a mathematics program of study for preservice grades K–8 teachers aligned with their needs as mathematics professionals? We are arguing that supplanting outmoded cultural regimes and lines of authority is extremely difficult. An outgrowth of this challenge is the politics of people frustrated with current conditions challenging professional competence. However, individual frustration masks the more pressing matters of institutional will, commitment, and unity of purpose.

As the two authors' personal perspectives suggest, a primary concern is developing an institutionalized means of supporting our partners in devoting time to issues of grades K–16 mathematics learning and to developing and maintaining cross-institutional and cross-expertise working relationships. As grant resources dry up, practice will have a strong inclination to settle back into the previously institutionalized norms: once you remove the external driver, systems often dampen back to their "fundamental frequencies."

A second concern is the extensive role that graduate students played in accomplishing SCALE work, particularly the work on preservice grades K–8 mathematics preparation. A concern with graduate students' level of involvement is that they remain at UW for a short period of time; we must thus develop a system to train and enculturate new graduate students consistently to replace those that leave. This does not say that graduate students' involvement is inappropriate. In fact, we believe the opposite. We could not have achieved much of KTI's and SCALE's accomplishments without the dedication of education and mathematics graduate students. Moreover, we hope that as graduate students with a strong involvement in SCALE activities move on to become faculty, they will bring with them a new perspective on the value of cross-institutional collaboration, thus extending SCALE's impact.

A final factor is the institutional constraints placed on faculty and, to a lesser extent, on staff and graduate students. As noted above, the relatively low emphasis placed on excellent instruction concerned participants in SCALE. Although SCALE may have achieved improvements in a person's practice and the value a person placed on those with different expertise, we need to understand these changes as occurring within the limitations of an institutional environment in which research is the primary focus. In the section that follows, we turn our attention to how we believe we might address these concerns by taking what we refer to as a systems approach to institutional change.

Moving Forward: A Systems Approach

One lesson from SCALE confirms a similar lesson found on a larger scale. In the recent evaluation of NSF's MSP program, Westat found that, even with well-positioned projects—those with multimillion-dollar budgets; dozens of organized, motivated, talented teams working with the best that research, content knowledge, and pedagogy have to offer; and operating for several years—the desired impact on students' test scores over the treatment period had not happened (Zhang et al. 2009). Certainly this disappoints the hopes that many have had going into these undertakings. However, this may not be a sign that these efforts were misguided, but rather that the adaptive, emergent challenges we face in these undertakings are greater than we had imagined, even though we assumed the challenges would be daunting.

One hypothesis that all this effort and research has led some of us to is the need to find a different organizational approach to span the multiorganizational jigsaw puzzle that constitutes the current cultural environment in which these efforts are embedded. It just might be that the current puzzle pieces cannot make a complete picture, no matter how we rotate and arrange them. We can formulate our hypothesis as follows:

> The quality and outcomes of professional learning in pre- and in-service mathematics and science will improve when key stakeholders of universities and school districts develop a common vision of high-quality content and pedagogy and through jointly sponsored instructional improvements create a lasting interface to coordinate material, human, social, and cyber resources to realize that vision.

Since much of this hypothesis is familiar in the NSF MSP context, we emphasize here that the statement's crucial component is that we must "… create a lasting interface to coordinate material, human, social, and cyber resources." Obviously, this has a great deal to do with institutionalization and sustainability. However, the approach that SCALE took at its inception was to treat the existing organizational structures as invariants in our work. That is, we treated the existing universities, colleges, departments, districts, subdistricts, schools, resource allocations, and authority lines as givens, and then worked from there. This, in turn, made SCALE strategies and actions particularly vulnerable to such things as funding, the cultural elements of these units, and turnover and policy changes as documented in the SCALE case study work.

We believe our hypothesis could be tested by attempting to create a stable organizational structure that (*a*) sponsors continuous improvement of professional learning over the long term, avoiding problems of temporary funding and shifting reform agendas; (*b*) has new authority and resources committed by leaders of the constituent departments, colleges, schools, and school districts, avoiding problems of intermittent support by decentralized entities; (*c*) can provide high-quality professional learning opportunities to expanded numbers of students, teachers, graduate students, faculty, staff, and administrators through strategic use of new and reallocated instructional resources; and (*d*) can foster the development of a professional learning community with distributed knowledge, practice, and skills.

The idea for this new type of organizational structure comes out of our experience and research over the past decade. As we alluded earlier, when both the KTI and SCALE work began, certain implicit assumptions worked against us. For example, we assumed that the district, math department, and school of education were all separate, distinct entities, and that responsibility, authority, and resource flow had to factor across those units in the ways they have in the past. In effect, we attempted to enact part of our hypothesis—creating a lasting interface among those entities—without considering the possibility that they were the wrong set of entities. It just might be, speaking metaphorically, that the math education equation with which we are working might not have a solution if those boundary conditions are a required constraint.

The details of such a new organizational structure will depend very much on context and involve many stakeholders and negotiations. The fundamental lesson learned from our experience and research is that providing new forms of professional learning to teachers and future teachers is not enough; rather, to be effective, we must apply professional learning reflexively to the entire system. Figure 10.2 partially represents this idea of a nested professional learning model. In this figure's inner circle, "K–16 Student's Domain," the instructor teaches students math and science content. In the next level, "K–16 Instructor's Domain," the instructor now learns about the previous level, "K–16 Student's Domain." In each successive layer, the former instructor learns about the previous level. The final level handles the instruction and learning collaboratively. In such a structure, the learner's nested role would extend throughout the system so that instructors, facilitators, and leaders all become learners in some circumstances and benefit from the professional learning. The key would be to create an economically viable model of such an organization—for example, by using interested graduate students who we first train and then provide professional learning, and, over time, by doing so at increasing levels in the nested diagram—that has the legal, resource, and authority foundations that allow for a solution to the mathematics education equation.

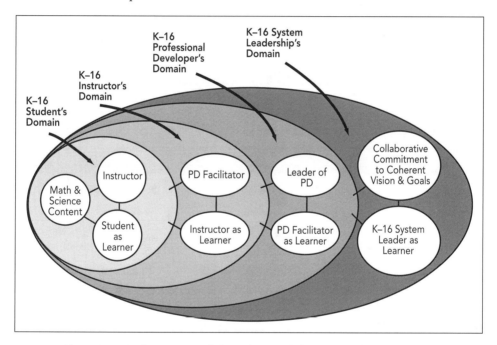

Fig. 10.2. Grades K–16 collaboration model, adapted from Ball and Cohen (2000) and Mumme and Seago (2002)

The SCALE partnership research work will culminate with their forthcoming book, *A Guide to Building Education Partnerships: Navigating Diverse Cultural Contexts to Turn Challenge into Promise* (Hora and Miller 2010). The findings discussed therein should be important not only for organizations interested in building partnerships in complex settings, but also for agencies and foundations interested in supporting such efforts. The lessons from this work will guide our future efforts as a loosely coupled learning community. We will work to create a new type of institutional partnership, working in what the book calls the "third space," as suggested in figure 10.3, from the book. The book comprehensively analyzes factors important for successful collaborative work, spanning such issues as understanding the culture, organization, and pivotal personalities of a partner institution; establishing formative and summative evaluation systems;

and encouraging individual and organizational learning. Certainly the SCALE work would have been more effective if at the beginning we had had the luxury of hindsight that this book offers.

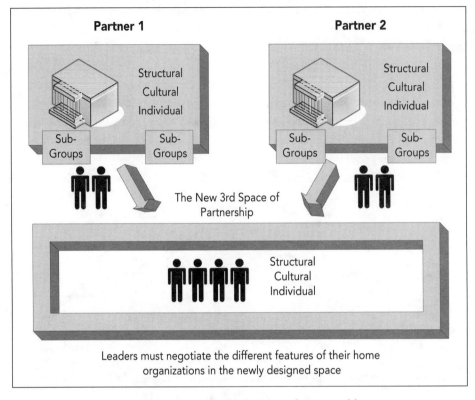

Fig. 10.3. The new third space of partnership

An important lesson is that if collaborative efforts focus on only a few of the kinds of factors identified in this work, then fundamental shifts in institutional practice will not happen. To use an analogy, sprinters can cover short distances very quickly, and, if properly trained, can extend their distance of effectiveness in unusual circumstances, assuming the extension is in tens or hundreds, but not thousands, of meters. However, sprinter training and experience alone is not proper preparation for marathon running. Likewise, handling only a few of the many aspects important to institutional collaboration as they come up is possible for reasonable people who have reasonable interpersonal and organizational skills. However, transformative institutional collaboration requires integrating all these aspects together. Attempting to integrate these aspects as they come up is not a manageable challenge. Anticipating the pitfalls that lie ahead in, speaking metaphorically, the marathon—or, perhaps more appropriately, the marathon combined with a steeplechase event—of transformative institutional change requires awareness and abstraction ahead of time.

REFERENCES

Ball, Deborah L., and David K. Cohen. *Challenges of Improving Instruction: A View From the Classroom.* Washington D.C.: Council of Basic Education, 2000.

Century, Jeanne R., and Abigail J. Levy. *Researching the Sustainability of Reform: Factors That Contribute to or Inhibit Program Coherence.* Newton, Mass.: Education Development Center, 2003.

Darling-Hammond, Linda, and Gary Sykes, eds. *Teaching as a Learning Profession.* San Francisco: Jossey-Bass Publishers, 1999.

Hall, Gene E., and Shirley M. Hord. *Implementing Change: Patterns, Principles, and Potholes.* 2nd ed. Boston: Pearson Education, 2006.

Hora, Matthew T., and Susan B. Millar. *A Final Case Study of SCALE Activities at UW—Madison: The Influence of Institutional Context on a K–20 STEM Education Change Initiative.* WCER Working Paper No. 2008-6. Madison: University of Wisconsin—Madison, 2008.

———. *A Guide to Building Education Partnerships: Navigating Diverse Cultural Contexts to Turn Challenge into Promise.* Sterling, Va.: Stylus Publishing, 2010.

———. *How to Diagnose and Manage Cultural Conflict in Educational Partnerships.* In preparation.

Lappan, Glenda, James T. Fey, William M. Fitzgerald, Susan N. Friel, and Elizabeth D. Phillips. *Connected Mathematics Series.* Glenview, Ill.: Prentice Hall, 2004.

Loucks-Horsley, Susan, Nancy Love, Katherine E. Stiles, Susan Mundry, and Peter W. Hewson. *Designing Professional Development for Teachers of Science and Mathematics.* 2nd ed. Thousand Oaks, Calif.: Corwin Press, 2003.

Mumme, Judith, and Nanette Seago. "Issues and Challenges in Facilitating Video-Cases for Mathematics Professional Development." Paper presented at the Annual Meeting of the American Educational Research Association, New Orleans, April 2002.

Senge, Peter M. *The Fifth Discipline: The Art and Practice of the Learning Organization.* New York: Doubleday, 1990.

Wilson, Suzanne M. *California Dreaming: Reforming Mathematics Education.* New Haven, Conn.: Yale University Press, 2003.

Zhang, Xiaodong, Joseph McInerney, Joy Frechtling, Joan Michie, John Wells, Atsushi Miyaoka, and Glenn Nyre. *Who Benefits? The Effect of STEM Faculty Engagement in MSP.* Rockville, Md.: Westat, 2009.

Reflection

William F. Tate

At no time in the past have we had a better opportunity, nor has it been more important, to create a viable knowledge base for effective research-based practices to improve student learning. The community can best create this knowledge base by linking the work and purposes of mathematics education researchers and practitioners who, by working together, can advance, respect, and benefit from each other's work. (Heid et al. 2006, p. 85)

THE RESEARCH-and-practice collaborations that the NCTM Research Committee called for are being implemented across the United States. The chapters in this volume represent a variety of approaches and organizational configurations that have linked research and practice in mathematics education. The NCTM Research Committee (Heid et al. 2006) noted that cultural differences between researchers and school practitioners could potentially inhibit reciprocal research-and-practice activities. In particular, the point of views or paradigmatic perspectives traditionally held by researchers and school practitioners differed on the relative importance of matters such as theory, analysis, and generalizability as well as interpreting findings as part of educational decision making. In addition, the NCTM Research Committee argued that methodological issues might also obstruct the link between research and practice. They were concerned that questions that interested practitioners be addressed as part of the research process. One concern with respect to research and practice is related. Is it feasible that research is sufficiently "on the ground" that it might guide midcourse corrections in a programmatic effort designed to advance mathematics teaching and learning? More generally, it is very important that research-and-practice collaborations inform participants of how well a programmatic effort is achieving its goals. Further, a need exists for these collaborations to generate evidence sufficiently robust and relevant to guide important decisions that school-based professionals make. The chapters in this volume represent examples of research-and-practice collaborations that, during their tenure of operation, dealt directly with using evidence to plan midcourse corrections, programmatic feedback strategies, and educational decision making.

The NCTM Research Committee (Heid et al. 2006) suggested that cultural and methodological difficulties contribute to the failure to build sustainable bridges between the communities of researchers and practitioners. The bridge linking the two communities is crucial to making advances in using research to guide organizational learning. In addition, opportunities for communication should concern themselves at some level with the following: (1) recording programmatic history, (2) offering feedback to practitioners and researchers, and (3) supplementing accountability measures. Research is a useful way to archive theories of action, decision making, and related outcomes. A wise colleague once remarked, "Ignorance of history is the mother of invention in education." Research provides an evidence trail for others to follow and learn. Appropriately communicated, research can provide instructive feedback to both practitioners and researchers. In my career, I have filled leadership roles as both a practitioner and researcher. In my capacity as a researcher, it has often been a challenge to understand fully the theory of action framing professional practice as well as the nature and quality of implementation. Working with educators to get a clear sense of these important aspects of practice is not trivial. In contrast, during my tenure as a leader of school practice in mathematics, I felt frustration with researchers who some in the organization viewed as asking the wrong questions while studying matters perceived as tangential to improved teaching and learning processes. This at times left a perception that research-based feedback was not central to the educational enterprise. Moreover, rarely did the researchers seek our thoughts, as leaders of practitioner groups, about the research process. This shortcoming in the feedback loop is a concern, because practitioners really need indicators and information to understand school processes and inputs better as part of information supplements to the outcome measures associated with most accountability systems.

The chapters in this volume represent a new beginning, where researchers, practitioners, and policymakers collaborated to form interdependent research and practice strategies in mathematics education. Although the collaborative efforts described in this volume largely reflect positive advances, many challenges still exist. Schmidt (2010, p. 51) succinctly captures a foundational divide:

> [M]ost likely a clash is inevitable between the research culture, where you constantly modify the plan on the basis of what you learn from the data, and the schooling culture, where plans are set quite far in advance and changes often cause conflicts or cannot be accommodated.

Schmidt and other authors in this volume make this argument for their research-and-practice endeavors. Why is this particular argument important? It speaks to the cultural and institutional challenge of building effective learning regimes in schools, where evidence and feedback link to modification as part of system corrective action. The past decade has seen the routine associated with research and evaluation become well established. Each year, school districts receive large repositories of data on their performance. This information, especially test scores, often arrives near the end of the academic school year. Moreover, process indicators to support a better understanding of the outcome measures are rare. A common theme among the chapters in this volume is a strong desire to disrupt this pattern. The cases presented here offer a vision of how research and practice might operate if supported long-term.

REFERENCES

Heid, M. Kathleen, James A. Middleton, Matthew Larson, Eric Gutstein, James T. Fey, Karen King, Marilyn E. Strutchens, and Harry Tunis. "The Challenge of Linking Research and Practice." *Journal for Research in Mathematics Education* 37 (March 2006): 76–89.

Schmidt, William H. "Building Bridges between Research and the Worlds of Policy and Practice: Lessons Learned from PROM/SE and TIMSS." In *Disrupting Tradition: Research and Practice Pathways in Mathematics Education*, edited by William F. Tate, Karen D. King, and Celia Rousseau Anderson, pp. 45–52. Reston, Va.: National Council of Teachers of Mathematics, 2010.